Coanda Effect

Coanda Effect

Flow Phenomenon and Applications

Noor A. Ahmed

CRC Press
Taylor & Francis Group
Boca Raton London New York

CRC Press is an imprint of the
Taylor & Francis Group, an **informa** business

CRC Press
Taylor & Francis Group
6000 Broken Sound Parkway NW, Suite 300
Boca Raton, FL 33487-2742

First issued in paperback 2021

© 2020 by Taylor & Francis Group, LLC
CRC Press is an imprint of Taylor & Francis Group, an Informa business

No claim to original U.S. Government works

Printed on acid-free paper

ISBN-13: 978-1-138-33915-6 (hbk)
ISBN-13: 978-1-03-209032-0 (pbk)

Visit the Taylor & Francis Web site at
http://www.taylorandfrancis.com

and the CRC Press Web site at
http://www.crcpress.com

To my wife and children

Contents

Preface

In the 1899 edition of the comedy magazine *Punch*, looking at the "coming century," an inventor asked, "Isn't there a clerk who can examine patents?" to which a boy replied, "Quite unnecessary, Sir. Everything that can be invented has been invented."

The reality remains: "Everything that can be invented, has not been invented."

The concept of the Coanda effect as a distinct flow phenomenon of high practical significance has existed for over a century. Its ability to be used as a flow control tool has been a great attraction to scientists and engineers. Flow control or manipulation of flow for a desired outcome continues to be instrumental in the development of many fluid mechanical devices underpinning the progress of human civilization. It is, therefore, not surprising that at the advent of the 4th industrial revolution, bringing with it astonishing advances in computer speeds, artificial intelligence, and manufacturing capabilities, the area of flow control and interest in Coanda effect has grown significantly. The prospect of implementing the Coanda effect for flow control solutions to wide-ranging problems, which would have been unthinkable in the past, is now looking more plausible. Yet the fluidic effect remains an enigma and its flow physics hard to decipher.

Interest in the Coanda effect originated from the belief that it would result in new and energy-efficient aircrafts by integrating lift and propulsive systems that would enable vertical take-off or require short runway length. But the weight penalty to carry the necessary equipment on board or the need to bleed off air from engines with extra ducting systems meant the concept could not be implemented with the intended energy-saving and performance efficiency. Researchers, therefore, have been exploring other non-aeronautical areas for application.

The list of areas where the Coanda effect can be applied is long and varied. It ranges from defense to health, from infrastructure to environment, and so forth. If the full potential of the Coanda effect is to be realized, it would require allocation of adequate resources, a multi-disciplinary approach, and collaborative efforts worldwide from researchers and engineers from various disciplines.

Unfortunately, however, there is no good book written on the Coanda effect that provides the basic information necessary. The present book is a humble attempt to fill this void, albeit partially. It is intended for scientists, engineers, and enthusiasts of varying backgrounds and levels of knowledge who are passionate about finding practical applications of the Coanda effect.

The Coanda effect can be thought of as a phenomenon of "flow adherence." This, of course, is a simplistic definition, but one that may be built upon and extended to provide greater insight. Flow adherence attributed to the Coanda effect is more than just flow attachment on a body and cannot be explained by conventional pressure gradient consideration of boundary layer theory alone. Additional flow characteristics such as enhanced turbulence level and entrainment accompany the Coanda effect. These features deserve special attention as they may result in increased noise levels that would be particularly important when it comes to designing Coanda devices and their implementation.

The book comprises of five chapters. In the first chapter, the reader is introduced to the basics of the Coanda effect through simple sketches and qualitative descriptions of the flow phenomenon. The conventional tools of investigation required to investigate the Coanda effect and other associated fluid flow phenomenon form the contents of the second chapter. The remaining chapters are devoted to research works related to Coanda effect applications. One of these chapters, namely Chapter 3, describes efforts on the aeronautical applications. Chapter 4 contains examples of non-aeronautical applications under the two subheadings of Industrial Applications and Environmental Applications. These two chapters are indicative of positive benefits that can be extracted from Coanda effect applications. The final chapter, Chapter 5, shows the negative effects of the occurrence of the Coanda effect in human blood flow and the airflow network, and how they can be avoided or detected early to ensure healthy performance of human organs.

It must be stressed that it is beyond the scope of this book to cover all the aspects of the Coanda effect or describe comprehensively the research works that are going on. The author has made no effort to claim the originality of the text presented, which has been motivated by the works of many others that have been appropriately referenced throughout the book. Often, for greater clarity and ease of understanding for the readers, the author has provided qualitative graphical descriptions of many of the results.

The book is of an introductory nature and should be viewed as a snapshot of ideas and possibilities. It is hoped that the materials presented will inspire the readers and appeal to their innovative talents for new breakthroughs. In the immortal words of Ulysses by Lord Tennison:

> "Come my friends,
> It is not too late to seek a newer world…
> For my purpose holds,
> To sail beyond the sunset…"

Happy reading!!

Noor A. Ahmed
1 January, 2019
Johannesburg

Author

Noor A. Ahmed has published extensively and delivered Keynote addresses at several international conferences on the application of aerodynamic concepts to find engineering solutions toward improving the quality of human living and comfort. He has developed advanced flow control and flow diagnostic techniques that have opened up new opportunities and wide areas of research that range from enhanced performance of aircraft to safe operation of wind turbines, from ventilation within enclosed spaces of buildings to environmental control systems of aircraft cabins, and so forth. The present book is a continuation of these efforts. It looks specifically at one flow phenomenon, called the "Coanda effect," which is not fully understood but has significant and diverse application potential. Starting with the basic concepts associated with the fluidic effect, the book provides examples of how the Coanda effect can affect aeronautical and non-aeronautical industries, contribute toward sustainable environment, and facilitate the healthy functioning of the human body in both developed and developing countries.

He earned his BSc (Hons) in Mechanical Engineering from Strathclyde University, UK, and PhD in Compressor Aerodynamics and Instrumentation from Cranfield University, UK. He worked as a Design Engineer for Kent Industrial Measurements, UK, and managed the Low Speed Compressor Test facility at Cranfield, UK. He was an Associate Professor and Plan Coordinator of Aerospace Engineering at the University of New South Wales, Australia, and is currently a Full Professor and Head of School at the Faculty of Engineering and Built Environment (FEBE) at the University of Johannesburg, South Africa.

He is a Chartered Engineer and a Fellow of the Institution of Mechanical Engineers, UK.

1

Basic Concepts

1.1 Historical Background

The beginning of the 20th century saw two momentous events occur in rapid succession; one happened in 1903, and took the world by storm, while the other happened in 1904, and almost went unnoticed. The two events, however, were necessary for the discovery and widespread popularity of what is now widely known as the "Coanda effect."

The event of 1903 marked the beginning of the conquest of air because on December 3 of that year, the first ever successful manned flight in human history by the Wright brothers took place [1]. This fired up people's imaginations, and accelerated developments in aviation began to occur. Within a few years, at the October 1910 Paris Salon de l'Aeronautique show, a young, 24-year-old engineer by the name of Henri Coanda was attracting the crowd's attention, displaying a new bi-plane aircraft he had designed to fly with a novel piston engine. But things did not go as planned. The aircraft was burned during its engine warm up before it could take off the ground [2]. Coanda was naturally disappointed. But there was a silver lining to the disaster when Coanda noticed while his engine burned, which, by 1934, led to the recognition of a new flow phenomenon that now bears his name as the "Coanda effect," a term coined by Theodore von Karman in his honor [2].

Coanda had observed that the burned gases which exhausted from the engine showed a tendency to remain very close to the fuselage. He was confident that he had discovered a new phenomenon and immediately began working on it to find practical applications. His initial objective was to deflect the exhaust gases, such as those of a radial piston engine away from the fuselage to protect the fuselage from getting burned. Later, he also experimented to apply the concept to turn flow occurring at corners of many devices, such as turning vanes of wind tunnels, thrust augmenters, pumps, and so forth. Coanda eventually managed the incredible feat of turning a flow through 180° deploying a series of deflecting surfaces, with each surface at a sharper angle to the previous one.

Historically speaking, however, Thomas Young was probably the first who had provided the first account of the flow phenomenon that produces the

Coanda effect in a lecture delivered to The Royal Society in 1800 [3], where he stated:

> The lateral pressure which urges the flame of a candle towards the stream of air from a blowpipe is probably exactly similar to that pressure which eases the inflection of a current of air near an obstacle. Mark the dimple which a slender stream of air makes on the surface of water. Bring a convex body into contact with the side of the stream and the place of the dimple will immediately show the current is deflected towards the body; and if the body be at liberty to move in every direction it will be urged towards the current.

In a paper [4] seventy years later, Osborne Reynolds observed a similar flow phenomenon when he discovered that a ball can be held in suspension by a jet of fluid. But it was Henri Coanda who had actually realized the practical importance of the effect and conducted detailed investigations to realize its potential. In 1934 he obtained a patent in France in which he described the effect as the "deviation of a plain jet of a fluid that penetrates another fluid in the vicinity of a convex wall." His two patents [5] filed in 1936 and granted in 1938 [Figure 1.1] also describe the Coanda effect quite clearly in the following text:

> The present invention relates to propelling devices in which there is produced a suction zone in front of the body in motion on which the propeller is mounted, this suction being such that the body in motion is propelled under the influence of the atmospheric pressure existing at the rear of the propeller.

The second historic event took place in 1904 when Ludwig Prandtl, at that time, a young, 29-year-old Professor of Mechanics at the Technical Institute of Hannover, Germany, delivered a ground-breaking paper [6] in which he explored the role of very small friction or vanishing viscosity in a fluid flow.

Viscosity is a major cause of drag produced on a body in motion, and the topic "drag" has received continued attention for a long time, from as far back as the times of Aristotle [7]. It was in the 18th century, nearly 2,000 years later that Navier and Stokes provided a mathematical formulation of a fluid motion that included the effects of viscosity. Unfortunately, the partial differential nature of the equations meant that they were difficult to use and remained unsolved until Prandtl paid attention to them.

Prior to Prandtl, every order-of-magnitude analysis to simplify the Navier–Stokes equations had resulted in potential flow equations suggesting that the flow was determined by the normal velocity component at the surface leaving no role for the viscous no-slip boundary condition, which at infinite Reynolds number would be exactly zero.

Prandtl also managed to eliminate, through a process of logical and intuitive deductions, the less significant terms from the Navier–Stokes equation

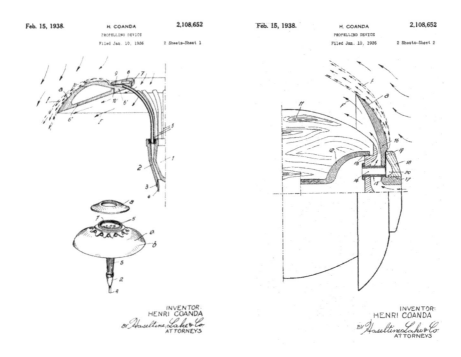

FIGURE 1.1
Henri Coanda's 1936 patent ([5].)

and reduced them into simpler forms with some significant differences. He argued that to satisfy the no-slip condition, there must be at least one term that would retain the effect of viscosity. He postulated that to compensate for the effect of vanishing viscosity, there must be an equivalent increase in the strain rate at the surface. Using this argument and from the corresponding simplified equations that he derived, he was able to deduce the viscous effects on a moving body in a stationary fluid and vice versa. He demonstrated that the viscous effects were dominant very close to the surface only. He also showed that from a point normal to the surface of the body, the static pressure remained constant within the viscous layer.

Prandtl's findings were highly significant. Now, for the first time, a viable and effective method for calculating the effects of viscosity near the surface of a moving body became possible. The scientific community did not take much notice of his work until a few years later, in 1908, when Blasius [8] proved the validity of Prandtl's hypothesis through physical experimentation on a flat plate. This was a turning point in the research on fluid flows which were until then studied without accounting for viscous effects and produced unrealistic results such as the D'Alembert paradox. The consequence was the development of what is now called "the boundary layer theory" and with that the beginning of modern fluid dynamics.

It is, therefore, clear that Coanda was an inventor who focused primarily on the applications of a new flow phenomenon rather than its detailed understanding. Prandtl, on the other hand, was an academic and a researcher who provided the theoretical basis for greater understanding of some of the underlying mechanisms of the fluid flow toward finding solutions of many of the problems of human society.

1.2 Simple Demonstrations of the Coanda Effect

In laymen's terms, "Coanda effect" is the tendency of a fluid to attach to, and flow around, solid surfaces. Generally, all jets, particularly plane or two-dimensional jets, display this flow phenomenon.

We can observe Coanda flow in some simple situations of our daily lives. Two such cases are shown in Figures 1.2 (a) and (b). In the first sketch,

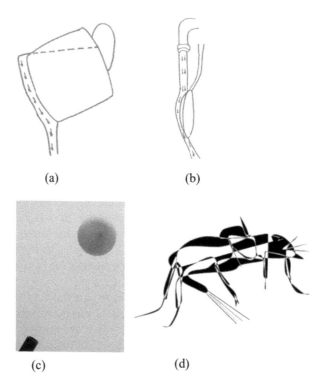

(a)   (b)

(c)   (d)

FIGURE 1.2
A jet of fluid is drawn toward a surface displaying the Coanda effect: (a) fluid attaches to the side of a cup during pouring; (b) water from a tap sticks to the side of a spoon; (c) a ball is held in suspension in air by a fluid jet (image after: Messiter, 1983 [9]); (d) a beetle sprays hot jet of streams over ants.

Figure 1.2 (a), when a liquid is slowly poured from a jug, it is seen to flow downwards hugging the surface of the jug. Similarly, in the second sketch, Figure 1.2 (b), when a spoon is held in the stream running down a water tap, the water flow is found to follow the surface contour of the spoon.

Another interesting example of the Coanda effect is shown in Figure 1.2 (c), which is based on the observation of Reynolds [9]. Reynolds had discovered that when a jet of air was thrown at an angle to a ball from a certain distance, the ball did not fall down but was held in suspension because of the jet that attached to the lower side of the ball. This was due to the Coanda effect. Gravity also played a part by preventing the ball from being blown away by the jet.

We can also observe the Coanda effect in action in nature as shown in Figure 1.2 (d) where a bombardier beetle can be seen exploiting it. Eisner and Aneshansley [10] describes the process in the following words:

> Bombardier beetles of the carabid subfamily Paussinae have a pair of flanges, diagnostic for the group, that project outward from the sides of the body. Behind each flange is a gland opening, from which the beetles discharge a hot, quinone-containing secretion when disturbed. The flanges are curved and grooved and serve as launching guides for anteriorly aimed ejections of secretion. Jets of fluid, on emergence from the gland openings, follow the curvature of the flanges and are thereby bent sharply in their trajectory and directed forward. The phenomenon is illustrative of the Coanda effect.

1.3 Manipulation of Coanda Flow

The significance of Coanda flow lies in the fact that it can be used as an effective flow control method. The potential list of areas where this method can be used for positive outcome is very long and varied, and ranges from aerospace industry to medical fields, from buildings to nuclear power plants, and so forth.

It is, therefore, important that we have a clear idea what is meant by flow control in the first instance and how the full potential of the Coanda flow phenomenon can be realized. For the latter, we need to have a good grasp of the flow physics of associated flows, which we will discuss later in Sections 1.4 and 1.5.

In the following sub-sections, we will attempt to introduce, in broad terms, the topic of flow control, its classification, methodologies, and intended outcomes.

1.3.1 Flow Control Definition

Flow control is a method to manipulate a fluid flow to achieve a desired outcome. It is, however, not a new concept. Flow control has been an integral part of nature and all living creatures since the time of creation. Animals, sea creatures, and birds have developed their bodies and organs to manipulate the flow around them. The bodies of cheetah or deer are streamlined to run fast. When fishes swim, they use fins to manipulate the water around them to navigate. When birds flap their wings, they are controlling flow around them to produce lift, or when they fly in a V-formation, they are manipulating the flow to conserve energy.

In prehistoric times, when *Homo sapiens* made spears, they basically designed them to manipulate the air flow so that the spears could travel faster and farther through air. Centuries ago, when the Persians invented windmills, they essentially created machines that would manipulate air flows to extract energy from wind.

Further examples of flow control can be found in abundance in nature. The progress of human civilization also has a lot to do with how flow control techniques have evolved and been utilized.

1.3.2 Role of Shear and Boundary Layers in Coanda Flow Control

A key feature of Coanda flow control is the manipulation of shear layers and boundary layers present in the flow. The introduction of boundary layer theory by Prandtl at the turn of the 20th century is relevant here because the theory can be regarded as the initiator of the modern era of flow control. The theory takes on added importance since, under many circumstances, the concepts of approximations applied to a flow near a boundary can also be adapted to shear flows outside the boundary.

One of the immediate consequences of the introduction of boundary layer theory was enabling the prediction of flow separation at a point on an arbitrarily shaped body caused by viscosity through energy dissipation.

Previously, flow separation could only be predicted on sharped edged or pointed bodies. The boundary layer theory opened up the possibility of controlling or delaying flow separation by arresting viscous growth within the boundary layer through sucking or blowing extra momentum to energize the fluid so that it remained attached to the body.

Delay or initiation of transition from laminar to turbulent boundary layer, suppression or amplification of turbulence in shear or boundary layer are the objectives of most flow control efforts. Often, most of these objectives are interdependent and complimentary. Judicious and selective choice of a particular approach are, therefore, required.

1.3.3 Flow Control Classification

Various definitions and classifications of flow control exist in the open literature [11]. The two most common ways of classifying flow control are based on whether the control is applied away from the surface or on the surface itself, or whether energy is expended or not to achieve flow control.

1.3.3.1 Contact or Non-Surface-Contact-Based

The flow chart depicting this form of classification is shown in Figure 1.3.

1.3.3.1.1 Control Applied away from the Surface

In this control classification, flow control is applied to outer flow fields of a surface. Examples include acoustic excitation of boundary layers on a surface, breaking up large eddies of shear layers into smaller ones, changing the turbulence of free streams and so forth.

1.3.3.1.2 Control Applied on a Surface

Contact surface flow control may involve changing the surface conditions of a surface, such as the shape, roughness, porosity, or temperature of the surface. Suction or injection of the boundary layer from slots on the surface or using boundary layer trips to control transition of flow from the laminar state are other examples of control applied to a surface.

1.3.3.2 Energy-Expenditure-Based

The second well-known way to classify a flow control is to consider whether energy is expended or not in implementing the control strategy. Based on this approach the control method employed is either passive or active, as shown in Figure 1.4.

1.3.3.2.1 Passive Flow Control

Passive flow control requires no energy input. The streamlining of a body such as an airfoil to improve aerodynamic efficiency is an example of passive control of a fluid. Since no demand is made for any additional energy input,

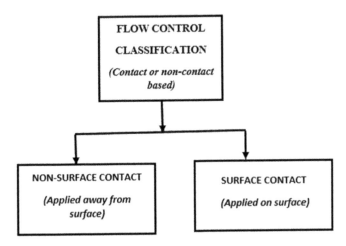

FIGURE 1.3
Classification of flow control based on non-surface or surface contact approach.

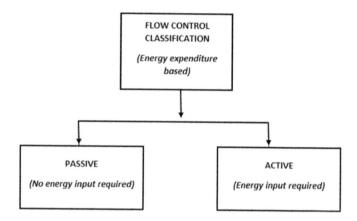

FIGURE 1.4
Classification of flow control based energy expenditure approach.

implementation of most passive control method is much easier to implement than active flow control.

1.3.3.2.2 Active Flow Control

Active flow control requires additional energy or a control loop, or both, and thus is more complex to implement than passive flow control. The concept of extra energy input is, therefore, the hallmark of the era of modern flow control. However, a very important requirement of all active flow control systems is that the manipulation or control of flows be done in the most cost-effective and energy-efficient manner. For that reason, various control loops may become essential and form integral parts of such systems. Hence, greater care is required in the design and management of the active flow control systems.

Despite these known complexities, active flow control may be preferred over passive systems in a large number of situations, particularly at off-design operating conditions, and to achieve higher efficiencies. Furthermore, the capability of switching on or off, as and apply when required, is another attraction of the active flow control systems.

1.3.4 Flow Control Methodologies

Various flow control methods or strategies are deployed to achieve the control objectives or outcomes mentioned earlier. Again, as before, depending on the methods deployed, the control methods may be grouped under three headings: transition control, separation control, and environment control. The flow chart is shown in Figure 1.5.

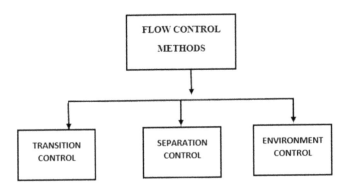

FIGURE 1.5
A simple flow chart of flow control methods.

Transition controls are employed to reduce transitional drag and reduce flow instabilities. Separation controls are employed to reduce viscous effects, thereby reducing drag. On an airfoil, this also leads to increases in the available angle of incidence and thereby contributes to higher lift.

It should, however, be pointed out that all the objectives and methods mentioned above rarely work in isolation, but are highly interdependent on each other.

1.3.5 Flow Control Outcomes

The outcomes desired from flow control can be grouped under three headings, namely aerodynamic efficiency, enhanced flow mixing, and reduced pollution. This is shown in Figure 1.6. Aerodynamic efficiency entails enhanced lift, reduced drag, or a combination of both. This is particularly important in the performance of wings. Greater flow mixing is a requirement

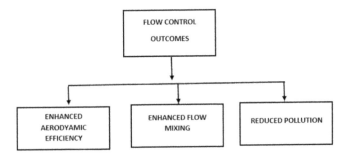

FIGURE 1.6
A simple flow chart of flow control outcomes.

of many industrial applications, such as in combustion of engines. Reduced pollution is desired for a better environment and life.

1.4 Understanding the Coanda Effect through Simple Sketches

We will now attempt to understand the flow physics of the Coanda effect. For that, we first need to know how a Coanda flow or a Coanda effect is created. From the examples presented in Section 1.2, it is clear that a jet flow and its subsequent movement over the surface of a wall hold the keys to understanding this complex flow mechanism.

We will examine the ensuing flow patterns qualitatively using simplified sketches and observe some of the basic flow features that give rise to the Coanda effect.

To keep matters simple, we will look at a two-dimensional incompressible jet of air as it flows into an otherwise stationary or still ambient air and note the changes in the flow in the absence or presence of a solid wall near the jet. Later, we will also look at the compressible effect.

In all the cases, we will assume the jet velocity to remain uniform and of constant magnitude.

1.4.1 Coanda Effect in Incompressible Flow

1.4.1.1 Jet Flow over a Straight Wall (i.e., Wall with No Curvature)

1.4.1.1.1 Wall at an Infinite Distance

Let us consider a straight wall that is located at an infinite distance away from a jet. In practice it means that there is no wall near the jet flow.

Under this scenario, the incompressible jet of air will draw the surrounding fluid or air molecules toward it and move them in the direction in which it is flowing. This will cause a decrease in the pressure of the air surrounding the jet, resulting in a flow pattern as shown in Figure 1.7. For the moment, we will refer to this process simply as "entrainment" and discuss the issue of entrainment in more detail in Section 1.4.2.

1.4.1.1.2 Wall with No Step

Let us now consider a straight wall without any step and place it parallel to the jet flow direction close to one side of the jet only.

Again, ambient air flow from the side away from the wall will continue to be drawn in a manner similar to that shown in Figure 1.7. But the process will be restricted by the nearer side of the wall because the air molecules that are removed from the region that lies between the jet and the wall become more difficult to be replenished by the ambient air.

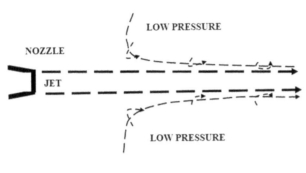

FIGURE 1.7
Flow patterns of a jet flow with no wall near it.

This will cause a larger pressure drop near the wall surface than on the side of jet where the wall is not present. The flow nearer the wall, as a consequence, will start to straighten and become more parallel to the wall as shown in Figure 1.8.

If the straight wall is now gradually brought closer to the jet, a further drop in air pressure will take place and at a particular distance eventually cause the jet to bend toward the wall as shown in Figure 1.9. In this scenario, the jet will suppress the boundary layer growth and the jet flow will continue by adhering to the wall.

This is how a Coanda flow is created and the adherence or hugging of the flow is called the Coanda effect.

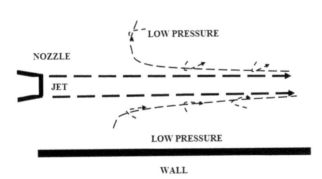

FIGURE 1.8
Flow pattern changes to a jet flow as a straight wall is brought closer to it.

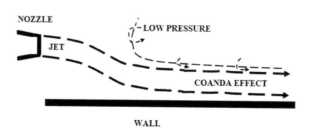

FIGURE 1.9
Flow pattern changes to a jet flow and Coanda effect is produced in the presence of a straight wall.

1.4.1.1.3 Wall with Step

Let us observe what happens when we add a step to the straight wall at its upstream or leading-edge end. The flow pattern in this case will be similar to that of Figure 1.9 except that it will contain a bubble or a recirculating region at the corner of the step, as shown in Figure 1.10.

Downstream of the bubble, the flow will continue its journey hugging the wall exhibiting the Coanda effect.

1.4.1.1.4 Inclined Wall with No Step

We will now consider the jet flow over an inclined wall without any step. The flow pattern will be similar to the flow observed in Figure 1.10 with the formation of a bubble at the junction of the nozzle and inclined wall, as shown in Figure 1.11.

1.4.1.1.5 Inclined Wall with Step

Addition of a step to the inclined wall of Figure 1.11 will not stop the Coanda effect taking place. Again, we will observe the formation of a recirculating

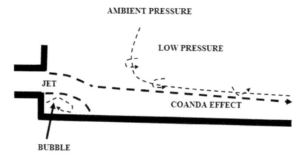

FIGURE 1.10
Flow pattern with Coanda effect in the presence of a straight but offset wall.

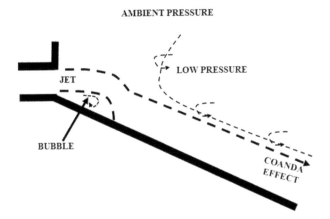

FIGURE 1.11
Flow pattern with Coanda effect in the presence of an inclined wall.

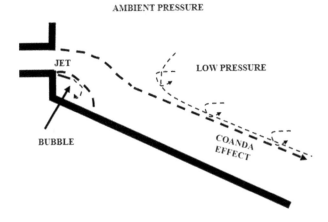

FIGURE 1.12
Flow pattern with Coanda effect in the presence of an inclined but offset wall.

zone or a bubble at the corner of the step and the flow patterns as shown in Figure 1.12.

1.4.1.2 Jet Flow over a Wall with Curvature

We will now look at a jet at a flow over a surface with curvature.

1.4.1.2.1 Concave Surface

Let us look at a jet flow over a concave surface.

The jet flow from the surrounding air will be drawn in from ambient flow and, similar to the cases of flow over a straight wall at a sufficient distance from the jet, the Coanda effect will be apparent in the flow patterns.

But the process will generate significant skin friction on the concave surface that will retard the movement of the jet flow. This will quickly create a rolling up of the flow on the surface and the formation of vortices giving rise to instabilities in the flow. These vortices that make the flow unstable on a concave surface are often called the Görtler vortices. A schematic of a flow depicting the Görtler vortices on a concave surface is shown in Figure 1.13.

1.4.1.2.2 *Convex Surface*

We will now look at the jet flow over a convex wall. The convex surface flow generally suffers lower losses when compared with the concave surface flows. Flow will also be much more stable. The adherence of a flow over a convex surface has often been referred to as Coanda flow in most literature.

However, as pointed out by Carpenter and Green [12], two types of flows are possible when a jet is introduced tangentially to a convex surface as typified by a circular cylinder. In one case, as reported also by Wille and Fernholz [13], the flow follows its journey adhering to the surface, showing the Coanda flow taking place, while in the other case the jet breaks away tangentially at the point of introduction. These two situations are depicted in Figure 1.14 (a) and (b), respectively.

So far, we have looked at jet flow applied externally to a solid body. The jet can also be introduced from inside the surface and exited through a slot. This may lead to the formation of Coanda flow on a convex surface. Figure 1.15 shows the formation of Coanda flow on either side of a Circular cylinder. Although not shown in the figure, the jet, as it bifurcates and turns sideways, allows for the formation of small re-circulating zones at the turning locations on either side of the cylinder.

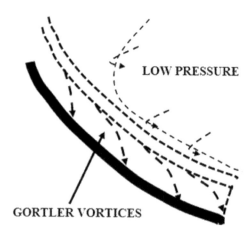

FIGURE 1.13
Schematic of Görtler vortices in a boundary layer on a concave wall.

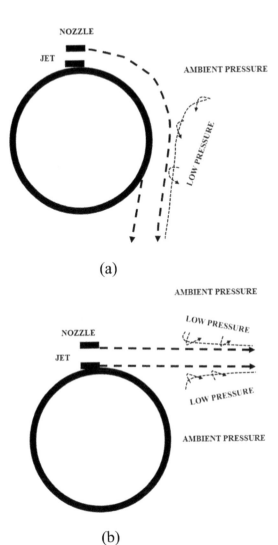

NOZZLE

JET

AMBIENT PRESSURE

LOW PRESSURE

(a)

AMBIENT PRESSURE

LOW PRESSURE

NOZZLE

JET

LOW PRESSURE

AMBIENT PRESSURE

(b)

FIGURE 1.14
Flow of a jet around a circular cylinder: (a) Normal Coanda flow; (b) Jet breakaway.

1.4.1.3 Jet Flow through a Channel or Tube

We will now consider a jet flow constrained by walls around it, such as jet flow through a channel or pipe.

1.4.1.3.1 Single Non-Bifurcating Channel or Tube

Let us first consider a jet flow through a single tube of constant diameter. If the jet velocity is assumed uniform and of constant magnitude, then in the absence of any disturbance, no Coanda effect is expected to form.

AMBIENT PRESSURE

BUBBLE

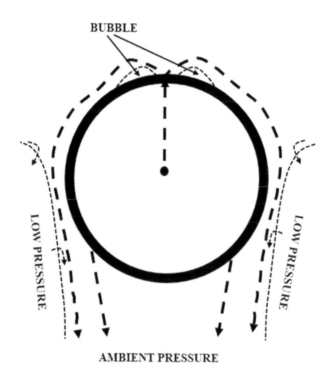

AMBIENT PRESSURE

FIGURE 1.15
Coanda flow on a convex surface from a jet exiting from inside the surface.

1.4.1.3.2 Single Bifurcating Channel or Tube

Let us consider the jet flow through the single tube bifurcating into two identical tubes of constant diameter. Coanda effect can also take place in such
bifurcating channels or tubes.

For any uniform flow without any disturbance, as shown in Figure 1.16 (a),
no Coanda flow will be formed. However, the presence of any non-uniformity in the parent configuration may result in a higher velocity in one of the
bifurcating tubes and the flow may proceed adhering to the side of this tube
as shown in Figure 1.16 (b).

We will describe the formation of Coanda flow in this way in greater detail
in Chapter 5.

1.4.1.4 Jet Flow in a Channel with Sudden Expansion

Let us now consider the case of a subsonic jet flow originating from a nozzle
that encounters an abrupt expansion. This can happen if the jet has to sud-
denly enter from a small configuration into a larger one.

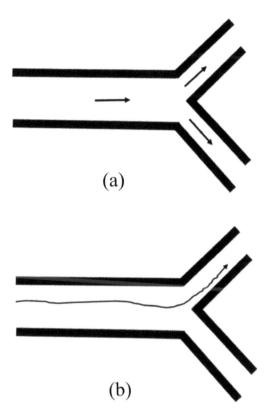

FIGURE 1.16
Flow bifurcation into a "Y" shape without any constriction in the parent tube: (a) uniform flow
and (b) non-uniform flow showing the Coanda effect.

Let us place a nozzle placed at the center of a large chamber that diverges
with equal magnitude of angle. Let us now introduce the jet through the
nozzle into an initially still and non-rotational fluid. Three possible scenarios
are possible as shown in Figures 1.17 (a)–(c).

In the first scenario, the jet continues to flow without any deviation. As it
flows, it creates low pressures of equal magnitude on either side and draws in
fluid toward its core as shown in Figure 1.17 (a). No Coanda flow is expected
to form in this case.

In the second scenario, the jet is deflected slightly to one side of the expanding wall. This is shown in Figure 1.17 (b). This time the jet produces low pressures zones of unequal magnitude on either side of it drawing in fluid from
its surroundings to its core. No Coanda flow will be created in this case.

In the third scenario, the jet becomes fully attached to one side of the
diverging wall. This will lead to the formation of a Coanda flow on this wall
as the jet flows adhering to this wall. Flow will be drawn in on the other side
of the jet as shown in Figure 1.17 (c).

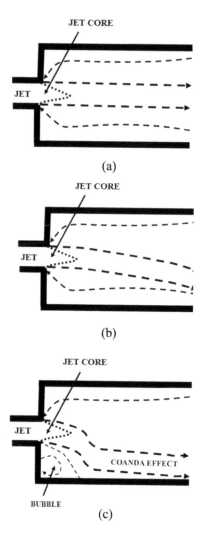

FIGURE 1.17
Schematic of three cases of a jet experiencing sudden expansion: (a) unattached non-deflected jet; (b) unattached deflected jet; (c) fully reattached jet on one side of the wall.

1.4.2 Coanda Effect in Compressible Flow

We have so far considered Coanda effect in incompressible and low-speed flows. Coanda flows found in compressible and high-speed flows are substantially different from incompressible flows. The density changes significantly alter the characteristics of the flow with the possibilities of formation of compressive or expansion waves and separation bubbles.

Just as in the incompressible case, the compressible jet will draw in ambient air from the surroundings, giving rise to a Coanda flow as demonstrated

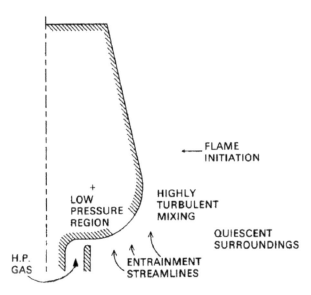

FIGURE 1.18
Schematic of Coanda flare in a compressible flow from the exit of a nozzle. (After: Gilchrist and Gregory-Smith, 1988 [14].)

by Gilchrist and Gregory Smith [14]. Some of the flow features are shown in Figure 1.18.

1.5 Basic Flows Associated with Creating the Coanda Effect

From the qualitative description presented in Section 4, it appears that the flow physics of a Coanda flow is dominated by free jet flow and boundary layer flow. Thus, greater knowledge of jet flow attachment and its detachment on various surfaces are essential if we are to eventually manipulate Coanda flow for a particular outcome.

For the sake of discussion, we will refer to an unbounded jet, that is, a jet that flows in the absence of any solid surface as a "free jet." This jet may be produced by a nozzle, which may be initially located external to and away from a wall before it flows over the wall. A "wall jet" may be produced internally from within a wall and ejected tangentially to its surface outside. We will often refer to bounded flows that flow in the presence of a wall as "wall flows."

Since a jet flow is a class of free shear flows, we will discuss it broadly under this heading. Our objective in this section is to gain an understanding of free shear and boundary layer flows.

1.5.1 Consideration of Free Shear and Boundary Layer Flows

Free shear flows have their origin from some form of surface upstream, such as a nozzle, a splitter plate, or a stationary or moving body. Such flows can be laminar or turbulent. All shear flows have the tendency to move from laminar to turbulent state quite rapidly, and once turbulent, they remain turbulent then onwards.

Boundary layer flows are flows over a surface requiring a no-slip velocity condition to be valid. These flows can also be laminar or turbulent. In nature, most commonly occurring free shear and boundary layers are turbulent.

In this section, we intend to gain some preliminary ideas about the origin and growth of these shear and boundary layer flows as they propagate. We also want to know how they transition from laminar to turbulent flows and the process whereby such flows remain attached to a body or suffer breakdown. These are complex issues, and despite decades of work, large gaps still exist in our understanding. We will, therefore, restrict ourselves mainly to some relevant aspects and provide descriptions that are essentially of introductory nature.

1.5.1.1 Velocity Profiles of Free Shear and Boundary Layer Flows

All free shear flows can be characterized by velocity profiles. Velocity profiles are a useful way of gathering information regarding any flow. These profiles can be represented graphically expressing the flow velocity as a function of the distance perpendicular to the direction of flow.

We will first look at the representation of free shear and boundary layer flows in terms of velocity profiles and to extract general information from such profiles to aid our understanding.

Some of the commonly encountered free shear flows and their velocity profiles are shown in Figures 1.19 (a)–(e). They include jet flow, mixing layer, wake flow, boundary layer flows on a flat plate, and inside a pipe. The first four of these flows can loosely be called external flows without or with the presence of a solid boundary, while the last is an internal flow within the confines of a solid boundary.

1.5.1.2 Laminar or Turbulent Flow from Velocity Profiles

Laminar flows propagate in an orderly manner in parallel layers without any disruption between the layers [15]. The flows do not contain eddies or swirls, lateral mixing, or cross-currents that are perpendicular to the direction of flow. Turbulent flows, on the other hand, are chaotic, produce eddies, swirls, and lateral mixing [16].

The effect of viscosity, if present in a flow, can be captured by velocity profiles. Generally, laminar velocity profiles can be represented by some mathematical functions, such as a sinusoidal function, that can be differentiated to give a measure of the skin friction at the wall directly. This is not possible

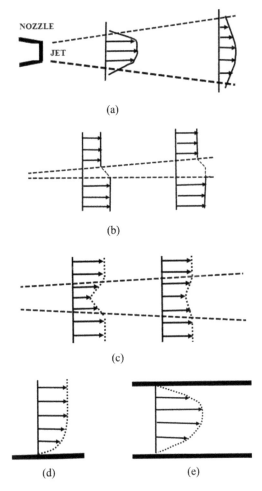

FIGURE 1.19
Velocity profiles of some shear and boundary layer flows: (a) free jet flows; (b) mixing layers; (c) wake flows of a non-lifting body; (d) boundary layer flows on a flat plate; and (e) boundary layer flow in a pipe.

for the turbulent flows and empirical correlations have to be used for the purpose. In fact, we can easily distinguish qualitatively between laminar or turbulent flows simply by looking at their velocity profiles. To demonstrate, we have provided laminar and turbulent velocity profiles in pipe flow in Figure 1.20.

1.5.1.3 Inflectional Velocity Profiles

All free shear and boundary layers are also characterized by inflectional mean velocity profiles. The inflectional attribute can be used to identify

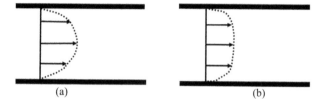

FIGURE 1.20
Velocity profiles in pipe flow: (a) laminar flow and (b) turbulent flow.

whether the flow will suffer inviscid instabilities or not. In such flow's viscosity acts as a damping mechanism and vortical induction plays the major role in producing instabilities.

To illustrate the process, let us consider three types of boundary layer velocity profiles, namely, a Blasius profile, a separation profile, and a flow reversal profile on a surface. These profiles are shown in Figures 1.21 (a)–(c).

In each of these figures, the first profile represents the normal profile in the streamwise direction while the second and the third profiles represent the first derivative and the second derivative, respectively, with respect to the direction normal to the surface (y-direction).

Generally, Rayleigh's "Point of inflection theorem" can be applied to an inviscid flow to determine whether a flow is stable or not. Rayleigh's theorem states:

> A necessary (but not sufficient) condition for inviscid instability is that the basic velocity profile $[U(y)]$ has a point of inflection [17].

In situations where Rayleigh's theorem is inconclusive in determining whether a flow is stable, Fjortoft's theorem can be applied. This theorem states:

> If yo is the position of a point of inflection, $[d^2U_\infty/dy^2=0$ in the basic profile $U_\infty(y)$ and $U_0= U_\infty(y_0)]$, then a necessary (but not sufficient condition for inviscid instability is that $(d^2U_\infty/dy^2) (U_\infty-U_0) < 0$, some where in the flow [17].

1.5.1.4 Displacement and Momentum Thickness from Velocity Profiles

Some very useful design parameters can further be obtained from the velocity profiles. These are displacement and momentum thicknesses.

Displacement thickness, δ_x, is defined as the distance by which the external flow is displaced by the decrease in velocity in the boundary layer and can be calculated using the following formula:

$$\delta_x = \int_0^\infty \left(1 - \frac{\rho u}{\rho_\infty U_\infty}\right) dy$$

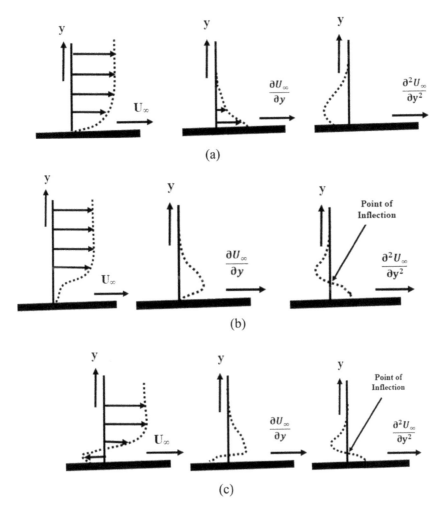

FIGURE 1.21
(a)–(c): Normal velocity profiles and their first and second derivatives with respect to the y-direction: (a) Blasius profile; (b) steady, two-dimensional separation profile; (c) flow reversal velocity profile.

In a similar manner, momentum thickness, θ_x, is defined as the distance to compensate for the reduction in momentum in the fluid caused by the boundary layer and can be obtained using the expression:

$$\theta_x = \int_0^\infty \frac{\rho u}{\rho_\infty U_\infty}\left(1 - \frac{u}{U_\infty}\right)dy$$

A graphical representation of the momentum and displacement thicknesses for a boundary length of δ is given in Figure 1.22.

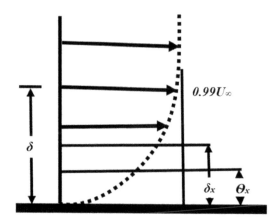

FIGURE 1.22
Relative magnitudes of displacement and momentum thicknesses for a boundary layer flow.

1.5.2 Free Jet Flow

We will now look at free jet flows in some detail.

A jet or a plume has excess momentum, which can be contrasted with a wake that has a deficit in momentum. The processes whereby excess momentum or momentum deficit are produced may vary. In the case of a free jet, for example, the excess momentum is produced by pressure gradient at source, whereas for a plume the excess momentum comes from the density gradients or buoyancy.

The structure of a free jet is highly dependent on the configuration from which it originates. Usually this configuration is a nozzle, as we have noted before. Consequently, the shape of the nozzle, whether it is round or elliptic, can have significant bearing on the nature of the boundary layers that form on the nozzle's inner walls that finally impact the characteristics of the free jet exiting the nozzle.

As an example, in the case of a thin jet issuing from a nozzle with laminar boundary layers, the jet produces coherent structures whereas the structures are less organized and random when the jet is thick and the nozzle boundary layer is turbulent.

1.5.2.1 Development of a Free Jet

A free jet, like all other free shear flows, as it develops in the absence of boundaries, becomes inhomogeneous and occupies an unbounded region of a large body of fluid that has either excess momentum or momentum deficit.

We will now consider the development of a subsonic free jet with the help of Figures 1.23 and 1.24.

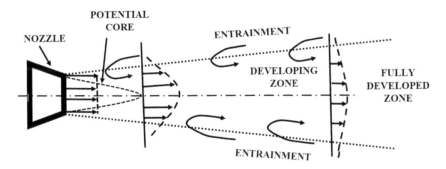

FIGURE 1.23
A schematic of plane laminar jet development.

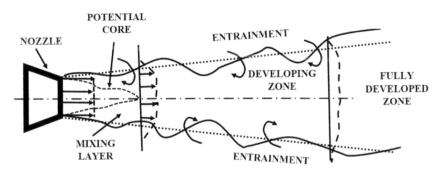

FIGURE 1.24
Schematic diagram of plane turbulent jet development.

1.5.2.1.1 Plane Laminar Jet Development

During the plane laminar jet development, the jet draws in the ambient fluid causing the flow rate across successive cross sections to change while the momentum of the flow starting at the orifice remains unchanged and is carried downstream.

A schematic of the typical development of a subsonic laminar free jet when it is introduced to an ambient atmosphere is shown in Figure 1.23.

1.5.2.1.2 Turbulent Jet Development

A plane turbulent development is shown in Figure 1.24.

The turbulent jet development is also similar to laminar jet development. The differences lie in the structure and the nature of their velocity profiles. The turbulent jet is made up of two plane mixing layers with a core of irrotational fluid in between. The turbulent jet develops as the two layers mix. The turbulent layer is chaotic and the size of eddies contained in it depends on the Reynolds number of the jet. The mean streamlines of the turbulent layer may not describe the actual flow patterns between the jet and the surrounding air. However, after mixing of the two layers, the jet becomes fully developed and becomes self-preserving.

1.5.2.2 Supersonic Jet Development

The supersonic jet flow development is, however, dependent on the relative magnitude of the nozzle exit and ambient pressure. Thus, there are three possibilities: one, an overexpanded supersonic jet when the nozzle jet pressure is higher, a matching supersonic jet when they are equal, or an underexpanded supersonic jet when the nozzle jet pressure is lower. Figures 1.25 and 1.26 show the cases of overexpanded and highly underexpanded jet.

For greater details about compressible flows, the reader can consult various textbooks [18, 19].

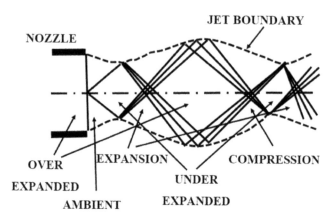

FIGURE 1.25
Schematic of various regions of an overexpanded supersonic jet.

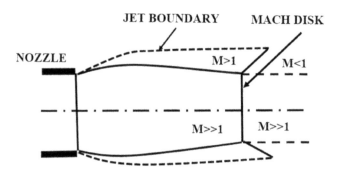

FIGURE 1.26
Schematic of highly underexpanded jet.

1.5.3 Flow Entrainment in a Jet

We have noted earlier that a free jet is created by pressure differences at the source. This means that it possesses velocity gradients that are caused by some upstream mechanism and are generally smoothed out by diffusion and convective deceleration. Since the jet discharges into an ambient fluid and is driven by its initial momentum, the jet spreads in a direction normal to the primary flow. The surrounding ambient fluid being stationary and irrotational leads to the creation of a low-pressure region. This draws in the irrotational ambient fluid that is incorporated into the turbulent region. The overall process whereby the free jet draws in the surrounding fluid is called "entrainment" as we have mentioned earlier in Section 1.4.1.

The lateral boundaries of a jet determine the spread rate of flow entrainment. A favorable pressure gradient, flow acceleration or heating increases the entrainment rate, whereas an adverse pressure gradient, or flow deceleration decreases it. The entrainment rate becomes significant, particularly when the flow is laminar or when there is not enough mixing. Under such circumstances, the radial spread of the mean velocity field and the decay of the mean velocity at the center-line in the downstream direction are hardly affected by any changes in the Reynolds number and are governed by the local velocity and length scales.

Studies by Fisher et al. [20] and Turner [21] found the entrainment coefficient of the jet to be lower than that of the plume. This implied that for the same local momentum flux, the dilution rate, or the rate at which irrotational fluid was incorporated into the turbulent flow was lower in jet than in plume.

Currently, however, no model exists that can satisfactorily explain all of the observed entrainment behaviors in jets. A simple expression for entrainment velocity, U_e, derived by Morton et al. [22] using the Taylor hypothesis, appears to have some appeal. Morten et al.'s formulation can be written as:

$$U_e = \alpha U_c,$$

where U_c is the time-averaged centerline velocity at that location, and α is the coefficient of entrainment whose numerical value depends on the local velocity profile and the chosen length scale.

The above equation of Morten et al. is found to be equally applicable to predict the behavior of plume. This may be because both jets and plumes are driven by the phenomenon of momentum excess, the former by their pressure gradient, and the latter by buoyancy. It is often convenient to compare the characteristics of both for a clearer understanding.

The effect of off-source heating, pressure gradients and other body forces on the entrainment of jets and plumes has been considered by Sreenivas and Prasad [23]. This is summarized in Figure 1.27.

Various models have been proposed to parameterize and explain entrainment in jets and plumes [24–27]. None of them appear capable of providing

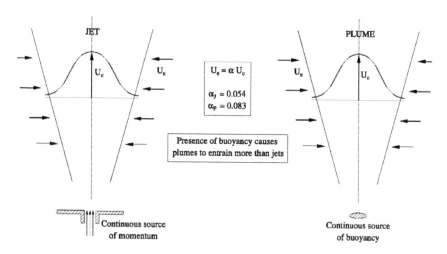

FIGURE 1.27
Effect of off-source heating, pressure gradients, and other body forces on jet and plume entrainment. (After: Sreenivas and Prasad, 2000 [23].)

the physical model that is required to explain the mechanism of entrainment in the near- and far-field adequately.

Based on the works of Gebhart et al. [28], near the origin of a shear flow, a vortex sheet separates the fluid containing momentum from the stationary ambient fluid. This vortex sheet travels downstream. It soon becomes unstable and rolls up forming a series of discrete vortices.

Vortex sheet roll-up is considered to be an instability process where the separation between these vortex structures is found to be proportional to the local width of the flow and the velocity profile. These formed eddies interact thereafter by a process of rolling and mutual induction of velocities. The rolling movements of these eddies entrap the irrotational ambient fluid that gets incorporated into the turbulent flow in a manner as shown in Figure 1.28.

The process of entrainment described appears to take place sequentially in three phases:

1. The induction phase [29] where the inducted fluid remains irrotational while it moves with the turbulent fluid, although still irrotational, forms a part of the moving turbulent.

2. Turbulent straining of the inducted fluid, which reduces its spatial scale to a small value with viscous diffusion dominating.

3. Infusion produced as a result of the viscous diffusion when the inducted fluid mixes at the molecular level with the turbulent flow.

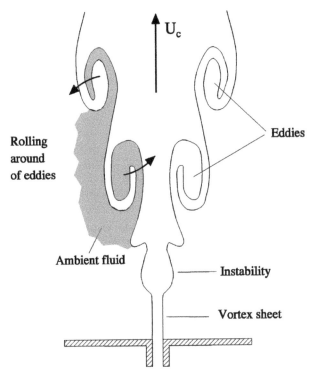

FIGURE 1.28
Axisymmetric jet in the near field; the shaded region indicates entrapped ambient fluid during the process of rolling around of eddies. (After: Sreenivas and Prasad, 2000 [23].)

Based on the above findings, Sreenivas and Prasad [23] concluded that the key step in the entrainment process is the induction phase or the rate at which ambient fluid is incorporated into the turbulent region. Works of Winant and Browand [30] and Brown and Roshko [31] provide further support to such eddy formation and rolling around in the near- and far-field of a mixing layer. Similar eddies are seen both in the near- and far-field of jets and plumes [32–37]; however, the mechanism of rolling of eddies around each other is seen only in the near field [34, 36, 37].

The mechanism of induction in jets and plumes appear to be different from mixing layers where the rolling around of eddies in mixing layers gives rise to the coalescence of eddies and the formation of a single larger eddy. Mixing layer entrainment also involves both the sweeping in of fluid by the induced velocity as well as the physically entrapping of the irrotational ambient fluid during rolling around of eddies. The jets and plumes, on the other hand, only involve the first mechanism. Thus, it is fair to conclude that in any shear flow, the entrainment process is heavily dependent on the induction velocity produced by the eddy structures, and any mechanism that hinders the induction process will subsequently affect the entrainment process.

Mungal and Hollingsworth [38] have found that eddies rolling around and coalescing in the far field of a buoyant jet is weak. A possible explanation suggested by them was that the induction mechanism in jets and plumes arose from a single eddy structure as sketched in Figure 1.29.

1.5.4 Instabilities in Shear Layers

In jet and flows transitioning from laminar to turbulent flows, instabilities are encountered and uncontrollable features may dominate the flow. Predicting such instabilities are not easy. We can, however, learn a lot by examining the velocity profile U_∞. Further treatments of the velocity profiles, such as their spatial first derivation or second derivation, can reveal additional features. The nature of these spatial velocity derivatives may also signify inflection

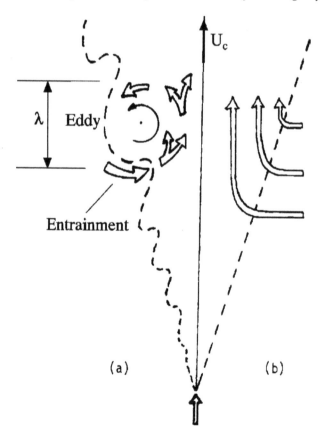

FIGURE 1.29
Jet-plume in the far field: (a) instantaneous entrained fluid motion with an eddy and instantaneous interface; (b) time-averaged entrained fluid motion and interface. (After: Hunt, 1994 [24].)

points of the velocity profiles, which in turn may indicate the onset of instability. Stability theories of flow are mathematically complex, and progress in this area has been slow.

1.5.4.1 Role of Viscosity

In an inviscid flow, to maintain equilibrium, the inertial and pressure forces need to stay in balance. A small change in velocity, from any source of disturbance, produces a pressure gradient. Thus, if there is acceleration due to a pressure gradient, then the addition of even a slight viscosity would retard the flow. For very low Reynolds number flow, this would produce a destabilizing effect.

Instabilities in jet profiles are, however, inviscid in nature and are caused by vortical induction. The function of viscosity in these flows is simply of damping nature. Jets are, therefore, characterized by inflectional velocity profiles similar to wakes or mixing layers. The jet velocity profiles are created when the jet exists from the surface of a body, such as a nozzle. Consequently, instabilities in jets generally have their origin at such exit surfaces.

1.5.4.2 Role of Reynolds Number and Mach Number

Both the Reynolds number and Mach number can have significant bearing on jet flows as they transition from laminar to turbulent flows. Jet flows undergo transition at much earlier low Reynolds numbers compared with wall flows. Usually the turbulence levels and flow entrainment are found to be lower in jet flows than in wall flows.

The transition process in jet flows is considered to be a consequence of the non-linear response of the laminar shear layer to random disturbances, such as the turbulence levels present in the jet stream. We will discuss transition issues further in Section 5.5 while discussing wall flows.

1.5.5 Wall Flow and Its Development

We will now look at wall flows. Compared to free shear flows, wall flows are more difficult to detect and characterize [39]. Despite a lot of work being conducted on wall flows, as evidenced by the vast amount of literature that is available in the public domain, the complexities associated with the nature of flow fields involved have exposed the limitations of existing methods of investigation. This has meant qualitative assessments and high dependence on flow visualization and correlation experiment results [40–46]. As a consequence, there exists a large divergence of opinions regarding the origin, structures and mechanisms associated with these flows. In this section, we will only highlight some of the relevant issues pertaining to wall flows. The interested reader is further encouraged to consult books and reviews [11, 17, 47–53] written on the topic for more details.

One way to get an appreciation of wall flows is to gain an understanding of the role that boundary layers play in such flows. It was realized, long before Prandtl, that the velocity at the interface between a stationary solid and a moving fluid must be zero. This is the "no-slip condition" that causes the shearing motion or produces the viscous effect with the consequence that velocity gradient must exist near every interface. Navier and Stokes were able to account for both the inertial and viscous terms present in a fluid flow in their momentum equations. But, as we have noted earlier in Section 1.1, that these equations proved too difficult to solve explicitly without resorting to some sort of drastic simplifications. In such circumstances, the Reynolds number, which indicates the relative importance of the inertia and viscous forces, was used to determine the extent to which these simplifying assumptions would hold for a particular condition. At a very low Reynolds number, mostly below the value of unity, inertia forces were ignored and Stokes law for drag was used to determine the drag for different shapes such as cylinder or sphere.

On the other hand, at a very high Reynolds number, that is, in the limit where the Reynolds number value tends toward infinity, viscous effects were ignored or assumed to be zero. Navier–Stokes equations were then reduced to the form of the Euler's equations. But these simplifications often gave results that did not correspond to what actually happened and produced the d'Alembert paradox that predicted zero drag on a circle or sphere in a stream, which was not borne out in physical experimentation results.

Prandtl's contribution was vital in providing the missing link between the theoretical and practical results. He argued that viscosity could never be totally ignored at a high Reynolds number and that viscous effects manifested in a thin layer above the solid boundary. This layer he had originally called the "transitional layer." But Blasius's subsequent reference to this layer as "boundary layer" became more popular and has since become universally accepted.

Within the boundary layer, as the Reynolds number increases, the boundary layer becomes thinner. For the no-slip condition to remain valid, high shear rates would be required to compensate for the decrease in the boundary layer. Since shear represents dissipation of momentum and energy, high shear rate implies a significant loss of total pressure within the boundary layer. Without knowing this loss, we cannot apply Bernoulli's equation to the velocity variations within the boundary layer since Bernoulli's principle is based on the assumption of constant total pressure.

Furthermore, a major problem arises when we seek to quantify the boundary layer thickness. Near the solid–fluid interface, the problem is easy enough to resolve because at the inner limit the flow velocity approaches a zero value almost in a linear fashion. But the situation is different for the outer layer as there is no upper limit of the boundary layer and the flow velocity approaches the free stream velocity asymptotically with increasing distance from the interface. So, the choice of the upper limit of the boundary is often arbitrary and is dictated by the circumstances under which it is considered. As a rule of thumb, as we have shown in Figure 1.22, the boundary length is taken to be the distance from the interface when the local velocity reaches approximately 99% of the free stream velocity.

Earlier, in Section 1.5.1.2, we observed velocity profiles to differ for laminar and turbulent flows. While the velocity profile for laminar profiles are somewhat easy to obtain, the same cannot be said for turbulent flows.

There appears to be eddies of various shapes and sizes that are superimposed on the overall horizontal turbulent flow over a surface. These eddies are very sensitive to factors such as buoyancy, temperature, or surface roughness. When buoyancy and temperature effects are negligible, the variation in velocity with height from a solid surface tend to follow somewhat of a logarithmic function. This conclusion has been arrived at from three different approaches. The first was the mixing length analysis by Taylor [54] for vorticity and later by Prandtl [55] for momentum, the second involved the dimensional analysis or similarity rule by von Karman [56], and the third was the asymptotic analysis by Millikan [57]. The essence of the log law or the law of the wall can be captured in the following expression:

$$U_\infty = \frac{U_*}{k} \ln\left(\frac{y-a}{y_0}\right)$$

where
k is the von Karman constant, generally taken as 0.4
a is the displacement length, taken to account for the extrapolation of the logarithmic value to a zero value on the surface
y_0 is the roughness length related to the size of the eddies generated on the surface
y is the height above the surface, and
U_* is the shear velocity or skin friction velocity obtained from the square root of the shear stress divided by the density of the fluid: $U_* = \sqrt{\tau/\rho}$

In a turbulent flow, it has often been the common practice to divide the flows above a surface into various regimes, such as viscous layer, an overlap layer, and an outer layer. The extent to which these regimes extend are usually described in terms of a new distance, y^+, which is a ratio of kinematic viscosity and skin friction, as expressed by:

$$y^+ = \frac{y}{d},$$

where d is the inner-layer length scale.

The viscous layer is further subdivided into two layers depending on the contribution of the turbulent shear stress. Between the region of $0 < y^+ < 5$, the viscous stress dominates, and the turbulence shear stress is essentially negligible; hence this region is called the viscous layer. Between $5 < y^+ \leq 30$, however, both viscous stress and turbulent shear stress become important, and this region exhibits a constant stress behavior and is often called the buffer layer. The region above $y^+ > 30$ is marked by the beginning of the outer layer where the turbulent shear stress begins to diminish and the regime within which the logarithmic law applies is called the overlap layer.

Using the notation of u^+ introduced by Spalding [58] to represent U/U_*, a curve of u^+ is drawn against y^+, and the various layers are shown in Figure 1.30.

The law of the wall has been questioned from time to time [11, 59–65]. We will not go into detailed discussion of these as they are beyond the scope of this book.

We will now look at the development of wall flows. A typical example of a straight wall flow is a two-dimensional flow on a flat plate. A flat plate by definition does not have any curvature and produces zero pressure gradient. As the flow moves from the leading to the trailing edge on the flat plate, there is initially a laminar boundary layer formed which is followed by a region of transition flow that eventually becomes turbulent due to some disturbance initiated at or near the surface of the wall.

The three flow regimes are shown in Figure 1.31.

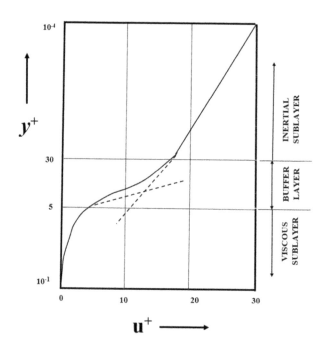

FIGURE 1.30
Various regimes of a turbulent wall boundary layer.

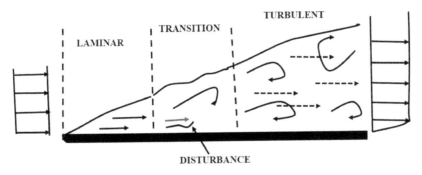

FIGURE 1.31
The three regimes of a fluid flow.

1.5.6 Transition in Boundary Layers

The causes of transition, be it in shear or wall flows from laminar to turbulent flows, are many and remain difficult to address. Laminar flows on a wall are extremely sensitive to the influence of any disturbance and the sources of such disturbances are many. Once transition has taken place, the flow becomes extremely unpredictable and the presence of various linear and non-linear mechanisms including Tollmein–Schlichting [66], Reynolds [67], centrifugal [68], crossflow [69–71], or Görtler [71–73] instabilities, may dictate the flow behavior of the boundary layers.

One way to explain the causes of the transition process is to use the streamline curvature and the rate of strain of the two processes [74, 75]. Generally, streamlines will curve due to the existence of pressure gradient in the direction across the streamlines. This pressure gradient causes centripetal force to be generated, which acts on the fluid particles and changes the direction of their flow paths. This leads to the balance between the pressure force and the centrifugal force to be changed and the flow may become unstable.

1.5.6.1 The Routes to Turbulence

The other approach involves stability considerations using the Morkovin approach [76–78], and we will now use this approach to get some idea of the transition process. To that end we have drawn up a flow chart presented in Figure 1.30. This flow chart is essentially a simplified representation of one of the versions of the "Morkovin Map," which appeared in the book by Panton [17]. The map is credited to Morkovin, who used it to describe the routes to transition of laminar flow to turbulent flow. But the fact that this map has been changed several times, and has the potential to change further in the future, underscores the many ambiguities and complexities that still remain unresolved. The significance of the map, however, lies in its ability to serve as a good guide to adopt some broad theoretical approaches that may involve linear, non-linear, or transient methods. It must, however, be stressed that none of these approaches can provide a complete picture of the flow transition process.

In Figure 1.32, the sources of disturbances have been grouped into two, one originating on the surface (as shown in Figure 1.31) and the rest are grouped under the heading "other disturbances." Very often, vigorous interactions between all sources of disturbances may take place making the process extremely difficult to analyze.

Roughness, curvature, vibration, or non-uniform temperature distributions are the main sources of surface disturbance [76, 79–83]. These give rise to instabilities in the laminar flow. The pathway to turbulence through instability is considered to be the natural way how the transition takes place.

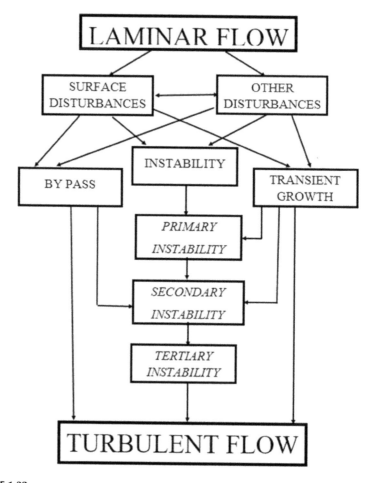

FIGURE 1.32
Flow chart of the route and processes associated with the transitioning of laminar flow to turbulent flow.

The primary instability in the flow chart of Figure 1.32, as the name implies, is the simplest form of instability in the boundary layer. It is linear and often viscous; a good example of primary instability is the disturbance originating from the Tollmein–Schilichting (TS) mode. The TS disturbances grow slowly and are dependent on particular Reynolds numbers for their amplification before they give rise to secondary instabilities. Often the TS disturbances fade away from the action of viscosity. The linear Orr–Sommerfield theory is often used to describe them.

In secondary instability, ∧-shaped patterns or hairpin vortices are seen to appear, which soon disintegrate completely into turbulence [84]. Various theories have been put forward [85–87] to explain the phenomenon. Once the primary instability attains a particular amplitude, the TS mode with the original laminar flow becomes periodic in nature with sub-harmonic growth. Secondary instability takes place, allowing faster inviscid growth, which eventually produces three-dimensional modes [88]. The final breakup of the three-dimensional modes is often called the tertiary instability, a precursor to turbulent flow on the wall.

Apart from the natural instability modes to turbulence, laminar flow can follow other routes. One route involves the free stream disturbances [89, 90] mixing with the laminar flow and proceeding directly to secondary instability. In this case, because of the absence of primary instability, it is called the "by-pass" route.

A third route to turbulence is through "transient growth" as shown in Figure 1.32. Plane Couette and Poiseuille flows fall under this category. These flows become unstable even at very low numbers [91] and have proved difficult to model using standard linear methods of analysis.

1.5.6.2 *Role of Viscosity and Reynolds Number*

We have already noted that a small change in velocity, from any source of disturbance, produces a pressure gradient and addition of small viscosity retarding the flow. Wall flows, both external and internal, are subject to the characteristics of the boundary layers formed on the wall surfaces. Under adverse pressure gradients, these boundary layers may also produce inflectional velocity profiles.

Generally, adverse pressure gradients encourage Tollmein–Schlichting and Mach instabilities. Convex surfaces, on the other hand, produce centrifugal instabilities [75] while inflectional cross flow velocity profiles or concave surfaces produce Görtler instabilities. The entrainment and turbulence levels of Coanda flows that are axisymmetric also increase noticeably through flow divergence [75, 92].

But unlike jet flows, viscosity plays a far greater role than just acting as a damping mechanism in wall flows. Thus, the instabilities produced are, to a greater extent, viscous in nature. Often the viscosity effect can also be destabilising to an otherwise stable profile. Viscous stresses generated near the wall have a time lag and can take some time to diffuse to the free stream. When the pressure force changes sign, the viscous force can combine with it to accelerate flow particles to larger amplitude giving rise to instabilities.

Viscous instabilities, however, are intrinsically stable. This makes the wall flow instabilities more difficult to control than jet flow instabilities. Instead of just looking at instabilities from an inviscid or viscous standpoint, they can also be examined by their linear response when subjected to some initial localized impulses. Huerre et al. [93] have used this approach and concluded that, in the absence of any flow separation, no mechanisms are available by which disturbances can propagate upstream of the boundary layers formed on a rigid wall surface. However, in the presence of a backward step, the re-circulating region caused by flow separation may facilitate the propagation of these disturbances upstream.

Similar to jet flows, wall flows are also significantly affected by variations in the Reynolds number and Mach number as they transition from laminar to turbulent flows. Wall flows undergo transition later and at much higher Reynolds numbers compared with jet flows. The Reynolds numbers have significant effect on the free-stream. With changes in the Reynolds numbers, the outer layer structure also changes. An accurate and reliable picture of the Reynolds number effect on the viscous sub-layer is still wanting. This is because the viscous sub-layer is very thin; this region poses significant difficulties to any physical mode of investigation, since most probe dimensions are greater than a viscous sub-layer thickness of $y^+ = 5$. This in turn leaves the results of any numerical investigation unvalidated.

1.5.7 Flow Separation

The final, important consideration in this chapter is flow separation on a surface. Viscosity creates friction on a wall flow that acts against the direction of the flow. This reduces the momentum of the flow and creates the adverse pressure gradient, $\partial p/\partial x$, in the boundary layer flow. The adverse pressure gradient is one of the causes of flow separation in a wall boundary layer flow.

In an incompressible flow, sharp edges may also lead to flow separation. In Figure 1.33, we have shown a horizontal flow that is impinging on a solid brick-like configuration. The flow, as it tries to negotiate around the corner, will separate from its edges as shown in Figure 1.33 (a). If, however, the frontal end is slightly rounded to a nose shape, the flow is better able to negotiate the corner. It may separate initially but may re-attach again, forming a region of zone of equilibrium pressure called a "separation bubble," as shown in Figure 1.33 (b). Further shaping of the body may eventually lead to what is called the "streamlining" of the body. The streamlining concept forms the basis of most lifting bodies, such as an airfoil, which is shown in Figure 1.33 (c). Here flow separates from the body at its trailing edge. Other situations when flow separation occurs is when there is a discontinuity in the surface,

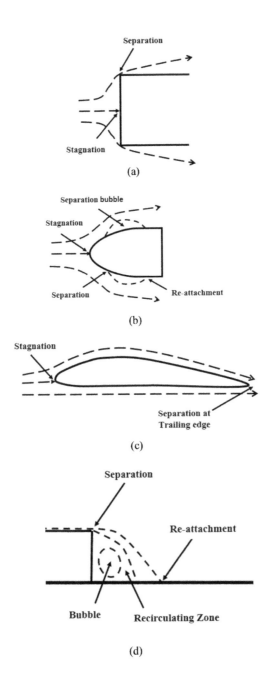

FIGURE 1.33
(a) Flow separation at sharp edges; (b) separation with the formation of a laminar separation bubble; (c) flow over a streamlined body with separation at its trailing edges; (d) separation at a step, bubble formation and re-attachment.

and a wake is formed downstream, such as at a step of a body as shown in Figure 1.33 (d).

The nature of boundary layers are important in flow separation. Laminar boundary layers are very susceptible to any disturbances, and are, therefore, more prone to flow separation than turbulent boundary layers. Turbulent layers can readily mix the low and high momentum fluid within the boundary layer. This helps slow down the adverse pressure gradient build up. This is one of the reasons why, in many practical scenarios, designers prefer to initiate turbulent boundary layers before adverse gradient is reached, and thereby prolong attached flow.

A flow in which the shear stress, τ, which is proportional to the velocity gradient, $\partial u/\partial x$, becomes zero, generally indicates the point of separation on a surface. The shear stress also becomes zero at an infinite distance from the wall. This means there exists a maximum stress within a flow which in turn suggest the existence of a point of inflection in a velocity profile. This inflection point can be more easily determined in a laminar flow than in a turbulent flow.

1.5.7.1 Laminar Boundary Layer Separation

Direct and inverse numerical methods can be employed to determine the point of separation for a steady two-dimensional laminar boundary layer. The direct method involves prescribing a known pressure gradient to solve the boundary layer equations [94]. This method, also known as the Goldstein's method, gives a singular solution where its shear stress becomes zero while the normal velocity, v, displacement thickness, δ_x, and longitudinal shear stress gradient, $\partial \tau/\partial x$ become infinite at the separation point. Inverse methods were developed to avoid the singularity and provide improved and easier solutions. In these methods [95, 96], generally the displacement thickness, δ_x, and the velocity distribution at the edge of the boundary layer $u_e(x)$ are prescribed to obtain the pressure gradient, $\partial p/\partial x$.

Separation bubbles formed in the laminar boundary layer separation can also be calculated using inverse methods [97, 98]. If, however, the shear stress becomes negative, there is back flow or flow reversal near the wall. This poses some difficulties in continuing with the boundary layer calculations. In such circumstances, some form of a boundary layer interaction method [99] can be used to obtain solutions.

1.5.7.2 Unsteady Boundary Layer Separation

In unsteady two-dimensional flow, even with vanishing shear stress on the wall or with flow reversal, flow separation may not occur. Rott [100], Sears [101], and Moor [102], working independently of one another, came to the conclusion that the unsteady boundary layer separation was akin to a situation when both the shear stress and velocity of separation vanished simultaneously over a moving wall, or when observed in frame of reference that moved with the separation point. This was further reinforced by the works of Sears and Telionis [103, 104]. But the difficulty of ascertaining the velocity with which the separation point moved made the approach difficult to locate the separation point when the wall was stationary. The problem was solved by Williams and Johnson [105, 106], who used the method of semi-similar solutions to transform a given unsteady flow into a steady flow for the moving wall flow and obtain solutions that were related to the unsteady case.

1.5.7.3 Turbulent Boundary Layer Separation

Turbulent boundary layer separation is also an area of immense practical importance. The structure of separated turbulent flows exhibit significant differences compared to attached flows. There are large scale structures in the separated shear layers that produce large pressure fluctuations, which can influence the back-flow zone, which has low velocity. Generally the back-flow is re-entrained into the outer flow region produces strong interactions between the pressure and velocity fluctuations. Although most separation occurs on walls with curvature, the effect of curvature is still not well understood. Other aspects such as imposed and self-induced unsteadiness can also affect separated layers significantly. These still remain areas of further studies. More detailed discussion regarding turbulent boundary layer separation can be found in text books and publications [107–109].

1.5.7.4 Three-Dimensional Boundary Layer Separation

We have so far considered only two-dimensional flows. Flow separation in three-dimensions are much more complex. There are some fundamental works done on various aspects of the topic [110–115]. Figure 1.34 shows flow separation on a three-dimensional body and the formation of a bubble and flow re-attachment. Figure 1.35 shows flow separation on a delta wing, where the flow separates along its edges, making the separation a controlled process. The separated flow rolls up into vortices which reattach back on the delta wing surface, giving rise to secondary flows and additional lift on the delta wing.

In three-dimensional boundary layers, we may see thin, viscous regions formed. This depends on whether there is lateral convergence or divergence of the flow. If the flow diverges away, say from an axisymmetric stagnation point, then the conservation of mass principle will ensure that there is thinning of the boundary. In a similar argument, the boundary layer will thicken if the steam lines converge to the stagnation point.

Longitudinal pressure gradient causes a plane flow to accelerate or decelerate. If transverse gradient is present, it adds another dimension to the layers and changes the direction of the streamlines to sideways. The static pressure

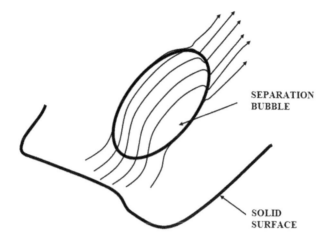

FIGURE 1.34
Flow separations in three-dimension leading to the formation of a separation bubble.

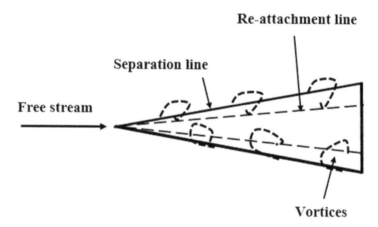

FIGURE 1.35
Controlled flow separation, formation of vortices and re-attachment on a delta wing.

within the boundary layer, however, does not change and remains constant. This has the effect of making the low velocity particles nearer to the surface of the boundary layer follow a tighter curve than the higher velocity particles at the edge of the boundary layer. The consequence is the initiation of a secondary flow.

The slight presence of transverse pressure gradient can make the streamline path highly curved, and the flow becomes three-dimensional. In general, separation in three-dimensional flow is a complex phenomenon and most of our knowledge has been formed based on flow visualization experiments and, therefore, can be characterized as qualitative in nature [116–118].

1.6 Concluding Remarks

From the introductory discussions we have presented in this chapter, it is apparent that jet flows and wall flows are complex and fundamentally different in character. The interactions of these two distinct flow mechanisms produce some mixed and complicated flow characteristics that form part of the Coanda flow phenomenon. Studies have shown the Coanda flows, in a global sense, are quieter and more stable than many conventional flows [119, 120]. Incorporating acoustics behavior adds further complexities to the study of Coanda flows, which will be evident in Chapter 4 and Chapter 5.

References

1. Anderson, J.D., (1997), *A History of Aerodynamics and Its Impact on Flying Machines*, Cambridge University Press.
2. https://en.wikipedia.org/wiki/Henri_Coandă
3. Young, T., (1800), Outlines of experiments and inquiries respecting sound and light, Quoted by J.L. Prichard, The Dawn of Aerodynamics, *Journal of Royal Aeronautical Society*, March 1957.
4. Reynolds, O., (1870), Suspension of a ball by a jet of fluid, *Proceedings of Manchester Literary and Philosophical Society*, vol. 9, p. 114.
5. https://patentimages.storage.googleapis.com/d9/67/6d/6cbdb5f33cc76e/US2108652.pdf
6. Prandtl, L., (1904) (in German), Über Flüssigkeitsbewegung bei sehr kleiner Reibung, in *Proceedings of 3rd International Mathematical Congress*, Heidelberg, Germany.
7. Ackroyd, J.A.D., Axcell, B.P., and Ruban, A.I., (2001), *Early Developments of Modern Aerodynamics*, The American Institute of Aeronauitcs and Astronauitics, Reston, VA, USA.
8. Blasius, H., (1908) (in German), Grenzschichten in Flüssigkeiten mit kleiner. Reibung, *Zeitschrift für Angewandte Mathematik und Physik*, vol. 56, pp. 1–37.
9. By Helmut Stettmaier – Own work, Public Domain, https://commons.wikimedia.org/w/index.php?curid=520630
10. Eisner, T., and Aneshansley, D.J., (1982), Spray aiming in bombardier beetles: Jet deflection by the Coanda effect, *Science*, vol. 215, issue 4528, pp. 83–85. doi:10.1126/science.215.4528.83
11. Gad-el-Haq, M., (2000), *Flow Control: Passive, Active and Reactive Flow Management*, Cambridge University Press, Cambridge, UK.
12. Carpenter, P.W., and Green, P.N., (1997), The aeroacoustics and aerodynamics of high speed Coanda devices, part 1: Conventional arrangement of exit nozzle and surface, *Journal of Sound and Vibration*, vol. 208, issue 5, pp. 777–801.
13. Wille, R., and Fernholz, H., (1965), Report of the first European mechanics colloquium, *Journal of Fluid Mechanics*, vol. 23, issue 4, pp. 801–819.

14. Gilchrist, A.R., and Gregory-Smith, D.G., (1988), Compressible Coanda wall jet: Predictions of jet structure and comparison with experiment, *International Journal of Heat and Fluid Flow*, vol. 9, No. 3, pp. 286–295.

15. Batchelor, G.K., (2000), *An Introduction to Fluid Mechanics*, Cambridge University Press.

16. Geankoplis, C.J., (2003), *Transport Processes and Separation Process Principles, (Includes Unit Operations)*, 4th edition, Prentice Hall Professional Technical Reference – Technology and Engineering.

17. Panton, R.L., (1996), *Incompressible Flow*, 2nd edition, Wiley and Sons.

18. Sapiro, A.H., (1953), *The Dynamics and Thermodynamics of Compressible Fluid Flow*, John Wiley & Sons, Hoboken, NJ.

19. Anderson, J.D. (2003), *Modern Compressible Flow with Historical Perspective*, 3rd edition, McGraw-Hill Publishers.

20. Fischer, H.B., List, E.J., Koh, R.C.Y., Imberger, J., and Brooks, N.H., (1979), *Mixing in Inland and Coastal Waters*, Academic Press, New York.

21. Turner, J.S., (1986), Turbulent entrainment: The development of the entrainment assumption, and its application to geophysical flows, *Journal of Fluid Mechanics*, vol. 173, pp. 431–437.

22. Morton, B.R., Taylor, G.I., and Turner, J.S., (1956), Turbulent gravitational convection from maintained and instantaneous sources, *Proceedings of the Royal Society, London, Series A*, vol. 234, issue 1196, pp. 1–23.

23. Sreenivas, K.R., and Prasad, A.K., (2000), Vortex-dynamics model for entrainment in jets and plumes, *Physics of Fluids*, vol. 12, issue 8, pp. 2101–2107.

24. Hunt, J.C.R., (1994), Atmospheric jets and plumes, in *Recent Research Advances in the Fluid Mechanics of Turbulent Jets and Plumes*, editors Davies, P.A., and Valente, M.J., Kluwer Academic, Dordrecht, the Netherlands, pp. 309–334.

25. Lumley, J.L., (1971), Explanation of Thermal Plume Growth Rate, *Physics of Fluids*, vol. 14, pp. 2537.

26. Priestly, C.H.B., and Ball, F.K., (1955), Continuous convection from an isolated source of heat, *Quarterly Journal of the Royal Meteorological Society*, vol. 81, issue 348, pp. 144–157.

27. Bhat, G.S., and Narasimha, R., (1996), Volumetrically heated jet: Large eddy structure and entrainment characteristics, *Journal of Fluid Mechanics*, vol. 325, pp. 303–330.

28. Gebhart, B., Jaluria, Y., Mahajan, R.M., and Sammakia, B, (1988), *Buoyancy Induced Flows and Transport*, Hemisphere Publishing Company, Washington, DC, USA.

29. Dimotakis, P.E., (1986), Two-dimensional shear-layer entrainment, *AIAA Journal*, vol. 24, issue 17, pp. 1791–1796.

30. Winant, C.D., and Browand, F.K., (1974), Vortex pairing: The mechanism of turbulent mixing layer growth at moderate Reynolds number, *Journal of Fluid Mechanics*, vol. 63, issue 2, pp. 237–255.

31. Brown, G.L., and Roshko, A., (1974), On density effects and large structures in turbulent mixing layers, *Journal of Fluid Mechanics*, vol. 64, issue 4, pp. 775–816.

32. Venkatakrishnan, L., Bhat, G.S., Prabhu, A., and Narasimha, R., (1988), Visualization studies of cloud-like flows, *Current Science*, vol. 74, pp. 597–606.

33. Dahm, W.A.J., and Dimotakis, P.E., (1987), Measurements of entrainment and mixing in turbulent jets, *AIAA Journal*, vol. 25, issue 9, pp. 1216–1223.

34. Yule, A.J., (1978), Large scale structure in the mixing layer of a round jet, *Journal of Fluid Mechanics*, vol. 89, pp. 413–432.

35. Dimotakis, P.E., Richard, C.M., and Papantoniou, D.A., (1983), Structure and dynamics of round turbulent jets, *Physics of Fluids*, vol. 26, pp. 3185–3192.

36. Rockwell, D.O., and Niccolls, W.O., (1972), Natural breakdown of planar jets, *ASME Journal of Basic Engineering*, vol. 94, pp. 720–728.

37. List, J.E., (1982), Turbulent jets and plumes, *Annual Review of Fluid Mechanics*, vol. 14, pp. 189–212.

38. Mungal, M.G., and Hollingsworth, D.K., (1989), Organized motion in a very high Reynolds number jet, *Physics of Fluids A: Fluid Dynamics*, vol. 1, pp. 1615–1623.

39. Liu, J.T.C., (1988), Contributions to the understanding of large scale coherent structures in developing free turbulent shear flows, in *Advances in Applied Mechanics*, editors Hutchinson, J.W., and Wu., T.Y., Academic Press, Boston, MA, USA, vol. 26, pp. 183–309.

40. Townsend, A.A., (1961), Equilibrium layers and wall turbulence, *Journal of Fluid Mechanics*, vol. 11, pp. 97–120.

41. Rotta, J.C., (1962), Turbulent boundary layers in incompressible fluids, *Progress in Aeronautical Science*, vol. 2, pp. 1–219.

42. Kline, S.J., Reynolds, W.C., Schraub, F.H., and Runstadler, (1967), The structure of turbulent boundary layer, *Journal of Fluid Mechanics*, vol. 30, issue 4, pp. 741–774.

43. Townsend, A.A., (1970), Entrainment and the structure of turbulent flow, *Journal of Fluid Mechanics*, vol. 41, pp. 13–46.

44. Offen, J.R., and Kline, S.J., (1974), Combined dye-streak and hydrogen-bubble visual observations of a turbulent boundary layer, *Journal of Fluid Mechanics*, vol. 62, issue 2, pp. 223–239.

45. Bakewell, H.P., and Lumley, J.L., (1976), Viscous sublayer and adjacent wall regions in a turbulent pipe flow, *Physics of Fluids*, vol. 23, pp. 1880–1889.

46. Willmarth, W.W., and Sharma, L.K., (1984), Study of turbulent structure with hot wires smaller than the viscous length, *Journal of Fluid Mechanics*, vol. 142, pp. 121–149.

47. Rosenhead, L., (1963), *Laminar Boundary Layers*, Clarendon Press, Oxford, UK.

48. Kovasznay, L.S.G., (1970), The turbulent boundary layers, *Annual Review of Fluid Mechanics*, vol. 2, pp. 95–112.

49. Laufer, J., (1975), New trends in experimental turbulent research, *Annual Review of Fluid Mechanics*, vol. 7, pp. 307–326.

50. Willmarth, W.W., (1975), Pressure fluctuations beneath turbulent boundary layers, *Annual Review of Fluid Mechanics*, vol. 7, pp. 13–37.

51. Cantwell, B.J., (1981), Organized motion in turbulent flow, *Annual Review of Fluid Mechanics*, vol. 13, pp. 457–515.

52. Blackwelder, R.F., and Swearingen, J.D. (1990), The role of inflectional velocity profiles in wall bounded flow, in *Near-Wall Turbulence*, editors Kline, S.J., and Afgan, N.H., Hemisphere Publishing, New York, pp. 268–288.

53. Robinson, S.K., (1991), Coherent motions in the turbulent boundary layers, *Annual Review of Fluid Mechanics*, vol. 23, pp. 601–639.

54. Taylor, G.I., (1915), Eddy motion in the atmosphere, *Philosophical Transaction of the Royal Society, London A*, vol. 215, pp. 1–26.

55. Prandtl, L (1925), Über die Ausgebildete Turbulenz (Investigations on turbulent flow), *Zeitschrift für Angewandte Mathematik und Mechanik*, vol. 5, pp. 136–139.

56. von Karman, T. (1930), Mechanische ähnlichkeit und turbulenz (Mechanical similitude and turbulence), in *Proceedings of the 3rd International Congress for Applied Mechanics, Stockholm, 24–29 August 1930*, editors Oseen, C.W., and Weibull, W. (Aktiebolaget Sveriges Litografiska Tryckerier, Stockholm, 1931), vol. 1, pp. 85–93; also NACA TM 611 (1931).
57. Millikan, C.B., (1939), A critical discussion of turbulent flows in channels and circular tubes, in *Proceedings of the of the 5th International Congress of Applied Mechanics*, editors Hartog, J.P.D., and Peters, H., Wiley, New York, pp. 386–392.
58. Spalding, D.B., (1961), A single formula for the law of the wall, *ASME Journal of Applied Mechanics*, vol. 28, pp. 455–457.
59. Barenblatt, G.I. (1979), *Similarity, Self-Similarity and Intermediate Hypothesis*, Plenum Press, New York.
60. Barenblatt, G.I. (1993), Scaling laws for fully developed turbulent shear flows, Part I, Basic hypothesis and analysis, *Journal of Fluid Mechanics*, vol. 248, pp. 513–520.
61. Barenblatt, G.I., and Prostokishin, V.M., (1993), Scaling laws for fully developed turbulent shear flows, Part II, Processing of experimental data, *Journal of Fluid Mechanics*, vol. 248, pp. 521–529.
62. Malkus, W.V.R. (1979), Turbulent velocity profiles from stability criteria, *Journal of Fluid Mechanics*, 90, pp. 521–39.
63. Long, R.R., and Chen, T.C. (1981), Experimental evidence of the existence of the meso-layer in turbulent systems, *Journal of Fluid Mechanics*, vol. 105, pp. 19–59.
64. George, W.K., Castillo, L., and Knecht, P., (1992), The zero pressure gradient turbulent boundary layer re-visited, in *13th Symposium on Turbulence*, editor Reed, X.B., University of Missouri, Missouri, USA.
65. George, W.K., and Castillo, L., (1997), Zero pressure gradient turbulent boundary layer, *Applied Mechanics Review*, vol. 50, issue 12, Part 1, pp. 689–729.
66. Schlichting, H., and Gersten, K., *Boundary Layer Theory*, 8th edition, Springer.
67. Squire, H.B., (1933), On the stability for three-dimensional disturbances of viscous fluid flow between parallel walls, *Proceedings of the Royal Society of London A*, vol. 142, pp. 621–628.
68. Billant, P., and Gallaire, F., (2005), Generalised Rayleigh criterion for non-axisymmetric centrifugal instabilities, *Journal of Fluid Mechanics*, vol. 542, pp. 365–379.
69. Stemmer, C., (2005) Transition investigations on hypersonic flat-plate boundary layer flows with chemical and thermal non-equilibrium, in *Proceedings of the 6th IUTAM Symposium on Laminar-Turbulent Transition*, Springer-Verlag, pp. 363–358.
70. Reed, H.L., and Saric, W.S., (1987), Stability and transition of three-dimensional flows, in *Proceedings of the 10th US National Congress of Applied Mechanics*, editor Lamb, J.P., ASME New York, pp. 457–468.
71. Reed, H.L., and Saric, W.S., (1989), Stability of three-dimensional boundary layers, *Annual Review of Fluid Mechanics*, vol. 21, pp. 235–287.
72. Stuart, J.T., (1963), Hydrodynamic stability, in *Laminar Boundary Layer Theory*, editor Rosenhead, L., Clarendon Press, Oxford, UK, pp. 492–579.

73. Di Prima, R.C., and Swinney, H.L., (1985), Instabilities and transition between concentric rotating cylinders, in *Hydrodynamic Instabilities and Transition to Turbulence*, editors Swinney, H.L., and Gollub, J.P., Springer-Verlag, Berlin, Germany, pp. 139–180.
74. Bradshaw, P., (1969), The analogy between streamline curvature and buoyancy in turbulent flow, *Journal of Fluid Mechanics*, vol. 36, pp. 177–191.
75. Bradshaw, P., (1979), Effect of streamline curvature on turbulent flow, AGARDograpgh no. 169, AGARD, 7 Rue Ancelle, 92200, Neuille Sur Seine, France.
76. Morkovin, M.V., (1969), Critical evaluation of transition from laminar to turbulent shear layers with emphasis on hypersonically travelling bodies, Air Force Flight Dynamics Laboratory Report No. AFFDL-TR-68-149, Wright-Patterson AFB, Ohio, USA.
77. Morkovin, M.V., (1984), Bypass transition to turbulence and research desiderata, I Transition in Turbines Symposium, NASA CP=2386, Washington, DC, USA.
78. Morkovin, M.V., (1988), Recent insights into instability and transition to turbulence in open flow systems, AIAA Paper no. 88-3675, New York.
79. Reshetko, E., (1976), Boundary layer stability and transition, *Annual Review of Fluid Mechanics*, vol. 8, pp. 311–349.
80. Reshetko, E., (1987), Stability and transition: How much do we know?, In *Proceedings of the 10th National Congress of Applied Mechanics*, editor Lamb, J.P., ASME New York, pp. 421–434.
81. Reshetko, E., (2001), Spatial theory of optimal disturbances in a circular pipe flow, *Physics of Fluids*, vol. 13, pp. 991.
82. Lee, C.B., and Wu, J.Z., (2008) Transition in wall-bounded flows, *Applied Mechanical Reviews*, vol. 61, issue 3, ASME, pp. 1–21.
83. Goldstein, M.E., and Hultgren, L.S., (1989), Boundary layer receptivity to long wave free stream disturbances, *Annual Reviews of Fluid Mechanics*, vol. 21, pp. 137–166.
84. Saric, W.S., and Thomas, A.S.W., (1984), Experiments on the subharmonic route to turbulence in boundary layers, In *Proceedings of the International Symposium*, Kyoto, Japan, September 5–10, 1983, (paper no. A85-24951 10-34), editor Tatsumi, T., Amsterdam, North Holland, 1984, pp. 117–122.
85. Craik, A.D.D., (1971), Non-linear resonant instability in boundary layers, *Journal of Fluid Mechanics*, vol. 50, pp. 393.
86. Herbert, T., (1988), Secondary Instability of Boundary Layers, *Annual Reviews of Fluid Mechanics*, vol. 20, p 487.
87. Saric, W.S., (1994), Low Speed boundary-layer transition experiments, *In:Transition: Experiments, Theory and Computations*, editors T.C. Corke, E.A. Erlebacher and M.Y. Hussaini, Oxford Press, UK.
88. Kelley, R.E., (1967), On the stability of an inviscid shear layer which is periodic in space and time, *Journal of Fluid Mechanics*, vol. 27, pp. 657.
89. Reshetko, E., (1994), Boundary layer instability, transition and control, AIAA Paper no. 94-0001.
90. Saric, W.S., (1994a), Gortler Vortices, *Annual Reviews of Fluid Mechanics*, vol. 20, p 379–409.
91. Tillmark, N., and Alfreddsson, H., (1992), Experiments on transition in plane Couette flow, *Journal of Fluid Mechanics*, vol. 235, p. 89.

92. Green, P.N., and Carpenter, P.W., (1983), Method of integral relations for curved compressible mixing layers with lateral divergence, in *Proceedings of the 3rd International Conference on Numerical Methods in Laminar and Turbulent Flows*, Seattle, USA, editors Taylor, C., Johnson, J.A., and Smith, W.R., Pineridge Press, Swansea, Wales, pp. 104–112.

93. Huerre, P., and Monkewitz, P.A., (1990), Local global instabilities in spatially developing flows, *Journal of Fluid Mechanics*, vol. 22, pp. 473–537.

94. Goldstein, S., (1948), On laminar boundary-layer flow near a position of separation, *Quarterly Journal of Applied Mathematics*, vol. 1, issue 1, p 43.

95. Cathedral, D., and Mangler, K.W., (1966), The integration of the two-dimensional laminar boundary-layer equations past the point of vanishing skin friction, *Journal of Fluid Mechanics*, vol. 26, pp. 163.

96. Carter, J.E., (1975), Inverse solutions for laminar boundary layer flows with separation and re-attachment, NASA TR R-447.

97. Cebeci, T., Keller, H.B., and Williams, J.C., (1979), Separating boundary layer flow calculations, *Journal of Computational Fluid Physics*, vol. 31, pp. 363–378.

98. Williams, J.C., (1977), Incompressible boundary layer separation, *Annual Reviews of Fluid Mechanics*, vol. 9, pp. 113–144.

99. Messiter, A.F., (1983), Boundary layer interaction theory, *ASME Journal of Applied Mechanics*, vol. 50, pp. 1104–1113.

100. Rott, N., (1956), Unsteady viscous flow in the vicinity of a stagnation point, *Quarterly Applied Mathematics*, vol. 13, pp. 444–451.

101. Sears, W.R., Some recent developments in airfoil theory, *Journal of Aeronautical Science*, vol. 23, pp. 490–499.

102. Moore, F.K., (1958), On the separation of unsteady laminar boundary layer, in *Boundary Layer Research*, editor Görtler, H., Springer-Verlag, Berlin, Germany, pp. 296–310.

103. Sears, W.R., and Telionis, D.P., (1972a), Unsteady boundary layer separation, in *Recent Research on Unsteady Boundary Layers*, editor, E.A. Eichelbrenner, Press de l' Universite Laval, Quebec, Canada, vol. 1, pp. 404–442.

104. Sears, W.R., and Telionis, D.P., (1972b), Two dimensional laminar boundary layer separation for unsteady flow or flow past moving walls, considering singularity due to bifurcating wake bubble, in *Recent Research on Unsteady Boundary Layers*, editor E.A. Eichelbrenner, Press de l' Universite Laval, Quebec, Canada, vol. 1, pp. 443–447.

105. Williams, J.C., and Johnson, W.D., (1974a), Semi-similar solutions to unsteady boundary layer flows including separation, *AIAA Journal*, vol. 12, pp. 1388–1393.

106. Williams, J.C., and Johnson, W.D., (1974b), Note on unsteady boundary layer separation, *AIAA Journal*, vol. 12, pp. 1427–1429.

107. Cebeci, T., and Bradshaw, P., (1977), *Momentum Transfer in Boundary Layers*, McGraw-Hill, New York.

108. Smith, A.M.O., (1977), Stratford's turbulent separation criteria for axially-symmetric flows, *Journal of Applied Mathematics and Physics*, vol. 28, pp. 920–939.

109. Simpson, R.L., (1989), Turbulent boundary layer separation, *Annual Review of Fluid Mechanics*, vol. 21, pp. 205–234.

110. Maskell, E.C., (1955), Flow separation in three-dimensions are much more complex, RAE Report Aero 2565, Royal Aircraft Establishment, UK.

111. Lighthill, M.J., (1963), Chapter 11, in *Laminar Boundary Layers*, editor Rosenhead, L., Clarendon Press, Oxford, UK.

112. Legendre, R., (1972), The Kutta-Jowkoski condition in three-dimensional flow, RA-1972-5, 241, 1972. RAE LT 1709.
113. Eichelbrenner, E.A., (1973), Three dimensional boundary layers, ARFM-5, no. 339.
114. Smith, J.H.B., (1975), A review of separation in three-dimensional flow, AGARD CP–168.
115. Küchemann, D., (1978), *The Aerodynamic Design of Aircraft*, Pergamon Press, Oxford, UK.
116. Tobak, M., and Peak, D.J., (1982), Topology of Three-Dimensional Separated Flows, *Annual Review of Fluid Mechanics*, vol. 14, pp 61–85.
117. Tani, I., (1977), History of boundary layer theory, *Annual Review of Fluid Mechanics*, vol. 9, Palo Alto, CA.
118. Perry, A.E., and Chong, M.S., (1987), A description of eddying motions and flow patterns using critical-point concepts, *Annual Reviews of Fluid Mechanics*, vol. 19, pp. 125–155, Palo Alto, CA.
119. Wilkins, J., Witheridge, R.E., Desty, D.H., Mason, J.T.M., and Newby, N., (1977), The design, development and performance of the Indair and the Mardair flares, in *Proceedings of the Ninth Annual offshore Technology Conference*, Houston, TX, pp. 123–130.
120. Fricker, N., Cullender, R.H., O'Brian, K., and Sutton, J.A., (1986), Coanda jet pumps: Facts and facilities, in *Proceedings of the International Gas Research Conference*, editor Cramer, T.L, held in Toronto, Canada, pp. 989–1003.

2

Tools of Investigation

2.1 Mathematical Treatment

The standard tools of investigation involve both theoretical and physical experimentation [1]. The theoretical approach requires application of the conservation laws in creating a mathematical description of a flow in terms of a set of equations. Direct analytical solutions of these equations are rare except for very simple cases. With the advent of high-speed computers, computational fluid dynamics has emerged as a powerful tool to solve equations numerically. These numerical methods, however, require, reliable and accurate data for validation and further advancement.

2.1.1 Conservation Equations

The Coanda effect can be studied theoretically by deriving the appropriate equations from the first principle and obtaining their numerical solutions. The primary sources of generating these equations are the conservation laws of mass, momentum, and energy. Starting with such numerical investigations first may be a good strategy from a design perspective because it can avoid many of the pitfalls of the trial and error approach or cost and time associated with physical experimentation. The main difficulty, however, lies in the fact that in every instance, the number of equations generated always falls far short of the number of unknowns that are present in these equations.

Further approximations and simplifications of the conservation equations are, therefore, inevitable if meaningful solutions are to be obtained. Careful examinations of the underlying assumptions and limitations of the conservation equations are essential before choosing procedures for their solution.

There are some good fluid mechanical books [2–9] that deal with the detailed derivation of the conservation equations and their practical significance. Such books are readily available in the open public domain and interested readers are encouraged to consult them for greater detail.

We will consider the conservation equations mainly in their final form and highlight the rationale that can be used to simplify them further for different scenarios. The objective is to reduce the number of unknowns with rational arguments before seeking solutions. This, we hope, will also help the reader understand the flow physics better and interpret more effectively the results obtained utilizing these equations.

We will often adopt Einstein's summation convention [10] to describe the equations for notational brevity.

2.1.1.1 Conservation of Mass Equation

The law of conservation of mass states that mass cannot be created or destroyed. For a fluid flow, a necessary assumption is that the fluid behaves like a continuum medium. This essentially transforms the conservation of mass equation into a continuity equation. A further assumption can be made that the fluid is non-reacting.

Then the following continuity equation can be obtained:

$$\frac{\partial \rho}{\partial t} + \frac{\partial}{\partial x_k}(\rho u_k) = 0 \tag{2.1}$$

It should be noted that since mass is a scalar quantity, only one equation can be constituted from the conservation of mass or continuity equation.

2.1.1.2 Conservation of Momentum Equation

The law of conservation of momentum is derived from a consideration of Newton's second law of motion which states that the rate of change of momentum can be directly linked to the forces generated. In an inertial frame of reference, the conservation of momentum can be thought of as a balance between the inertia, pressure, viscous, and body forces.

The final form of the equation when expressed in the unit of force per unit volume can be written as:

$$\rho\left(\frac{\partial u_i}{\partial t} + u_k \frac{\partial u_i}{\partial x_k}\right) = \frac{\partial \Sigma_{ki}}{\partial x_k}(\rho u_k) + \rho g_i \tag{2.2}$$

Momentum being a vector quantity, equations such as the one derived above gives rise to three component equations, one each for the x, y, and z directions, respectively.

2.1.1.3 Conservation of Energy Equation

The law of conservation of energy states that in non-relativistic motions, energy can neither be created nor destroyed. It can be expressed in different forms.

The thermal form of the energy equation can be written as:

$$\rho\left(\frac{\partial e}{\partial t} + u_k \frac{\partial e}{\partial x_k}\right) = -\frac{\partial q_k}{\partial x_k} + \sum_{ki} \frac{\partial \sum u_i}{\partial x_k} \tag{2.3}$$

Here again, energy being a scalar quantity, the law of conservation of energy furnishes one equation only.

2.1.2 Reducing the Number of Unknowns

The three conservation laws of mass, momentum, and energy outlined in Equations (2.1), (2.2), and (2.3) provide us with five differential equations as depicted in equations. But in these equations there are a total of 17 unknowns, namely ρ, u_i, $\sum_{ki}$, e and q_k. In other words, there are more unknowns than equations. Clearly, therefore, the equations do not constitute a determinate set, and no solutions can be obtained by using them.

To make the equations determinate, we need to impose several justifiable assumptions that may help to reduce the number of unknowns.

Below we describe some of the assumptions that have been adopted and the rationales behind them:

- No body couples. This causes the stress tensor to become symmetric, reducing the number of its independent components to six.
- Thermodynamic equilibrium. This allows justification of linear relationships to be considered between the stress-rate of strain and heat-flux temperature.
- Newtonian fluid. This implies that a linear relationship between the stress tensor and the symmetric part of the deformation tensor, or the rate of strain tensor, can be adopted.
- Isotropic fluid. This signifies that the constant of proportionality is scalar and helps to reduce the constants of proportionality dramatically, from over 81 to only two [7].
- Fourier fluid. This suggests that the conduction part of the heat flux vector can be linearly related to the temperature gradient. Along with the isotropy assumption mentioned above, this Fourier Fluid assumption implies that the constant of proportionality in this relation is just a single scalar.
- Fluid behaves like an ideal gas. This enables the application of the gas law relating the density, pressure, and temperature to be applied:

$$p = \rho RT \tag{2.4}$$

- Stokes hypothesis. This helps link the first and second coefficients of viscosity (μ and λ) using the relationship:

$$\lambda + \frac{2}{3}\mu = 0 \tag{2.5}$$

- There is no radiative heat transfer. This makes another set of relationships available to link spatial temperature gradients to heat transfer:

$$q_i = -K_f \frac{\partial T}{\partial x_i} \tag{2.6}$$

With the above assumptions, the momentum and energy equations can be simplified further, while the continuity equation will remain unchanged.

The complete set of conservation Equations (2.1), (2.2), and (2.3) can now be re-written as:

$$\frac{\partial \rho}{\partial t} + \frac{\partial}{\partial x_k}\left(\rho u_k\right) = 0 \tag{2.7}$$

$$\rho\left(\frac{\partial u_i}{\partial t} + u_k \frac{\partial u_i}{\partial x_k}\right) = -\frac{\partial p}{\partial x_i} + \frac{\partial}{\partial x_k}\left[\mu\left(\frac{\partial u_i}{\partial x_k} + \frac{\partial u_k}{\partial x_i}\right) + \partial_{ki}\lambda \frac{\partial u_{kj}}{\partial x_j}\right] + \rho g_i \tag{2.8}$$

where, expressed as per unit volume,

$$\rho\left(\frac{\partial u_i}{\partial t} + u_k \frac{\partial u_i}{\partial x_k}\right) = \text{inertia forces}$$

$$-\frac{\partial p}{\partial x_i} = \text{pressure forces}$$

$$\frac{\partial}{\partial x_k}\left[\mu\left(\frac{\partial u_i}{\partial x_k} + \frac{\partial u_k}{\partial x_i} + \partial_{ki}\lambda \frac{\partial u_k}{\partial x_i}\right)\right] = \text{viscous forces,}$$

$$\rho g_i = \text{external forces}$$

$$\rho C_v\left(\frac{\partial T}{\partial t} + u_k \frac{\partial T}{\partial x_k}\right) = \frac{\partial}{\partial x_k}\left(K_f \frac{\partial T}{\partial x_k}\right) - p\frac{\partial u_k}{\partial x_k} + \Phi \tag{2.9}$$

where,

$$\Phi = \frac{1}{2}\mu\left(\frac{\partial u_i}{\partial x_k} + \frac{\partial u_k}{\partial x_i}\right)^2 + \lambda\left(\frac{\partial u_j}{\partial x_j}\right)^2 \tag{2.10}$$

Equations (2.7), (2.8), and (2.9) now contain six unknowns, namely, ρ, u_i, p, and T. The equation of state provides the sixth equation relating ρ, p, and T. So, in theory, using these equations and specifying a sufficient number of boundary conditions, under most circumstances, and for most fluids, numerical solutions of both laminar and turbulent flows can be obtained.

The above equations can also be used for compressible and high-speed flows, as long as the Mach number of the flow is less than 2. The Mach number is a non-dimensional number that represents the ratio of inertia force and compressible force or, put in another way, how fast the flow is moving compared to the speed of sound in the flow medium. As a rule of thumb, the equations are usable if the Mach number of the flow is less than 2; in other words, in flows that are highly subsonic or supersonic but produce weak shock waves. Greater care has to be exercised if the flow is transonic, that is when the flow's Mach number equals one or is close to one.

2.1.3 Well-Posed Incompressible Equations

Further simplifications can be made to the above equations, namely to Equations (2.8) and (2.9), if the flow is incompressible, that is, when compressibility effects are negligible. In general, this is the case when the flow's Mach number is less than 0.3. The density ρ in this flow, then, is a constant or virtually remains a constant. This would reduce the number of dependable variables to five. Since we have five equations from the conservation laws, the set of equations becomes determinate, and is, therefore, amenable to solution.

Furthermore, it is worth noting that the convection of enthalpy in the incompressible flow is balanced by heat conduction and viscous dissipation. So, Panton [11] has suggested that in the energy Equation (2.6), the specific heat at constant volume, C_v, should be corrected by replacing it with the specific heat at constant pressure, C_p.

The simplified incompressible flow equations can now be written as:

$$\frac{\partial u_k}{\partial x_k} = 0 \tag{2.11}$$

$$\rho\left(\frac{\partial u_i}{\partial t} + u_k \frac{\partial u_i}{\partial x_k}\right) = -\frac{\partial p}{\partial x_i} + \frac{\partial}{\partial x_k}\left[\mu\left(\frac{\partial u_i}{\partial x_k} + \frac{\partial u_k}{\partial x_i} + \right)\right] + \rho g_i \tag{2.12}$$

$$\rho C_p\left(\frac{\partial T}{\partial t} + u_k \frac{\partial T}{\partial x_k}\right) = \frac{\partial}{\partial x_k}\left(K_f \frac{\partial T}{\partial x_k}\right) + \Phi_{incompressible} \tag{2.13}$$

Equations (2.11)–(2.13) are valid for both non-turbulent and turbulent flows. When centrifugal, gravitational, electromagnetic, or other forces are absent or insignificant, the non-dimensional Reynolds number, which signifies the relative magnitude of the inertia and viscous forces generated in a flow, can be used to characterize whether a flow is non-turbulent or turbulent.

2.1.3.1 Non-Turbulent Flows

For cases where the Reynolds number are very small (Re ≪ 1), that is, the inertial forces are very small compared to the viscous forces, the inertial forces can be neglected in the momentum equations. Such flows can be considered to non-turbulent or laminar. On the other hand, when the Reynolds numbers are very high, that is, when the inertial forces are much larger than the viscous forces, the flow becomes turbulent and transient in nature.

2.1.3.2 Turbulent Flows

Most flows of practical significance are turbulent. Except for simple cases, turbulent flows have always proved extremely difficult, if not impossible, to solve. This is because the dependable variables in turbulent flows are random functions of space and time for which no general stochastic solutions exist. Some non-stochastic solutions may be possible by adopting the method of "Reynolds averaging," named after Osborne Reynolds, who first proposed [12] the approach in the late 19th century, and this approach remains relevant to this day.

The Reynolds method involves ensemble averaging, which can be temporal or spatial of the dependable variables where the conservation equations are written for the various moments of the fluctuations about their mean. The possibility of achieving a solution increases if the flow field can be assumed stationary or homogenous. This makes the time derivative of any statistical quantity become zero and the temporal functions independent of space.

To understand the process better, let us consider a Newtonian turbulent flow with constant density and viscosity in which velocity, pressure, and temperature fluctuate randomly. In this instance, there are five random variables, namely u_i, p, and T. Reynolds proposed the decomposition of each random variable to its mean and fluctuating components, so that:

$$u_i = \overline{U}_i + u_i';$$
$$p = \overline{P} + p'$$

and

$$T = \overline{T} + T'$$

where the symbols $\overline{U}_i$, $\overline{P}$, and $\overline{T}$ (with bars) and u_i' and p' (with primes) represent the average and fluctuating quantities of instantaneous values of u_i, p, and T, respectively.

Substituting the instantaneous values of u_i, p, and T in Equations (2.11), (2.12), and (2.13) with their corresponding average and fluctuating components, the Reynolds averaged turbulent flow equations become:

$$\frac{\partial \overline{U}_k}{\partial x_k} = 0 \tag{2.14}$$

$$\rho \frac{\partial \bar{U}_k}{\partial t} + \bar{U}_k \frac{\partial \bar{U}_k}{\partial x_k} = \frac{\partial \bar{P}}{\partial x_i} + \frac{\partial}{\partial x_k}\left(\mu \frac{\partial \bar{U}_i}{\partial x_k} - \overline{\rho u_i u_k} \right) + \overline{\rho g_i} \tag{2.15}$$

$$\rho C_p \left(\frac{\partial \bar{T}}{\partial t} + u_k \frac{\partial \bar{T}}{\partial x_k} \right) = \frac{\partial}{\partial x_k}\left(K_f \frac{\partial \bar{T}}{\partial x_k} \right) + \Phi_{\text{incompressible}} \tag{2.16}$$

Now there are five equations and 11 unknowns in the Equations (2.14), (2.15), and (2.16). Apart from the viscous stresses and pressures, there are additional turbulent stresses, called the Reynolds stresses, $-\overline{\rho u_i u_k}$, present in the momentum equations. To close the system of equations, the Reynolds stress terms are generally expressed as a function of the mean flow without any reference to the fluctuating components. This computational procedure is known as turbulence modelling [13].

For most engineering applications, detailed information of the turbulent fluctuations are not required and different turbulence models have been devised to seek solutions for different situations.

2.1.4 "Karman Approach" for Incompressible and Compressible Flows

If a flow is two-dimensional or axisymmetric in the mean, then a possible is one that was developed by von Karman [14], which resulted in what is now known as the Karman integral momentum equations. Pohlhausen [15] was able to obtain results for two-dimensional laminar flows that established the approach to be a viable one. The approach soon became quite effective for both laminar and turbulent flows [16, 17] irrespective of whether the flow was Newtonian or not.

As mentioned in Section 2.1.3.2, for turbulent flows, the decomposition method of Reynolds' offers a practical way to consider the additional turbulent stresses. Let us examine the Karman equations before integration, which is given below:

$$\rho \frac{\partial \bar{U}}{\partial t} + \rho \bar{U} \frac{\partial \bar{U}}{\partial x} + \rho \bar{V} \frac{\partial \bar{U}}{\partial y} = -\frac{\partial \bar{P}_\infty}{\partial x} + \rho \frac{\partial}{\partial x}\left(\overline{v^2} - \overline{u^2} \right) + \frac{\partial \tau_x}{\partial y} \tag{2.17}$$

The last two terms of the equation, namely, $\rho \frac{\partial}{\partial x}\left(\overline{v^2} - \overline{u^2} \right)$ and $\frac{\partial \tau_x}{\partial y}$, deserve some attention:

- The term $\rho \frac{\partial}{\partial x}\left(\overline{v^2} - \overline{u^2} \right)$ is difficult, and often, impossible to integrate. An order of analysis carried out by Hinze [18] shows the term could be neglected in the context of boundary layer approximation.

- When the term $\rho \frac{\partial}{\partial x}\left(\overline{v^2} - \overline{u^2} \right)$ is neglected, the solution obtained by integrating Equation (2.17) makes the Karman integral method of first order magnitude.

- The term $\dfrac{\partial \tau_x}{\partial y}$ can be obtained using the mean streamwise velocity, $\bar{U}$. In the absence of suction or injection, the wall shear stress term can be obtained from $\tau_w = \mu \dfrac{\partial \bar{U}}{\partial y}\bigg]_{y=0}$

We will now consider Karman's approach further. In an external uniform flow, Karman observed similarity in the velocity distributions within the boundary layer at different sections. This feature allowed him to treat the partial differential equations of the boundary layer as ordinary differential equations, and consequently enabled him to combine the continuity and momentum equations into a single equation.

Integral equations can also be obtained in a similar manner from kinetic and thermal energy differential equations.

For a two-dimensional incompressible flow, Karman's integral equation can be written in the following form [19]:

$$\frac{C_f}{2} = \frac{1}{U_\infty^2}\frac{\partial}{\partial t}(U_\infty \delta_x) + \frac{\partial \delta_x}{\partial x} - \frac{\rho_w v_w}{\rho_\infty v_\infty} + \delta_x\left[\left(2 + \frac{\partial \delta_x}{\partial \theta}\right)\frac{1}{U_\infty}\frac{\partial U_\infty}{\partial x} + \frac{1}{\rho_\infty}\frac{\partial \rho_\infty}{\partial x}\right] \quad (2.18)$$

where,

$$C_f = \frac{\tau_w}{\frac{1}{2}\rho_\infty U_\infty^2} \quad (2.19)$$

$$\tau_w = \mu \frac{\partial U}{\partial y}\bigg]_{y=0} + \mu \frac{\partial V}{\partial x}\bigg]_{y=0} \quad (2.20)$$

The second term of Equation (2.20) can be neglected when injection or suction are absent.

An important feature of Karman's equation is the appearance of the displacement thickness, δ_x, and momentum thickness, θ_x, terms. These terms carry special physical significance (see Section 1.4.1.4 of Chapter 1), and can provide useful information in practical applications.

To solve Karman's integral equation, however, we have to know in advance or prescribe velocity profiles of the flow. In a laminar flow, appropriate velocity profiles are easy to define. From these velocity profiles, the velocity gradients normal to the wall can be obtained and shear stress at the wall determined for laminar flows.

Velocity profiles for turbulent flow are much more difficult to define. Various velocity profiles and empirical relationships have been proposed for turbulent flows with varying degrees of success, and application of Karman's equations have been mostly limited to simpler cases. An empirical correlation, due to Ludwig and Tillman [20], is generally used to evaluate the skin friction coefficient in turbulent flow, is given as:

$$C_f = \frac{0.246}{10^{0.678H} R_\theta^{0.268}}$$ (2.21)

where H is the shape parameter $\left(= \frac{\delta_x}{\theta} \right)$.

A rapid rise in the value of the shape parameter, H, (H > 2) can be used to predict the onset of flow separation on a body. For more accurate determination of the point of flow separation, a modified shape parameter, H*, proposed by Head [21] is often used. Head hypothesized that the controlling factor in the growth of a turbulent boundary layer is dictated by the entrainment of the fluid into the boundary layer from the mainstream flow adjacent to it, and he derived the following relationship:

$$\frac{1}{U_e} \frac{d}{dx} \left[U_e \left(\delta - \delta_x \right) \right] = F \left(H_* \right)$$ (2.22)

where $H_* = \frac{\left(\delta - \delta_x \right)}{\theta}$

Head further argued that if a one-parameter family of turbulent velocity did exist, then there should also exist a unique relationship between the shape parameter, H and his modified parameter, H*, giving:

$$H_* = G \left(H \right)$$ (2.23)

Head went on to determine the functions F and G that were used to determine the separation value of H* to be equal to 3.34 [9].

2.1.5 Equations for Fluid Flow near a Wall with Little or No Curvature

Since the Coanda effect is concerned with free jet flow and flow near a surface, it is important to obtain the relevant wall equations that represent the amount of fluid that crosses the wall in a flow per unit time at any one instant. These equations can be derived from the global conservation equations we have presented earlier. Since a solid surface is the cause of vorticity generations in a flow, the resulting equations can also be used to determine vorticity of the flow.

Let us consider an incompressible fluid flow over a stationary straight wall, that is, on a wall that has a small or zero curvature.

In this flow, let v_w represent, at $y = 0$, the velocity normal to the wall. The positive or negative value of v_w signifies injection or suction, respectively. If the body forces are neglected, then the instantaneous momentum equations

for both laminar and turbulent flows in the streamwise, normal, and spanwise directions at $y = 0$ can be written as:

$$\rho v_w \frac{\partial u}{\partial y}\Bigg]_{y=0} + \frac{\partial p}{\partial x}\Bigg]_{y=0} - \frac{\partial \mu}{\partial y}\Bigg]_{y=0} \frac{\partial u}{\partial y}\Bigg]_{y=0} = \mu \frac{\partial^2 u}{\partial y^2}\Bigg]_{y=0} \qquad (2.24)$$

$$\frac{\partial p}{\partial y}\Bigg]_{y=0} = \mu \frac{\partial^2 v}{\partial y^2}\Bigg]_{y=0} \qquad (2.25)$$

$$\rho v_w \frac{\partial w}{\partial y}\Bigg]_{y=0} + \frac{\partial p}{\partial z}\Bigg]_{y=0} - \frac{\partial \mu}{\partial y}\Bigg]_{y=0} \frac{\partial w}{\partial y}\Bigg]_{y=0} = \mu \frac{\partial^2 w}{\partial y^2}\Bigg]_{y=0} \qquad (2.26)$$

The above equations can be useful in determining the streamwise and spanwise vorticity as well.

2.1.6 Equations for Fluid Flow near a Wall with Curvature

Coanda flows occurring on walls with curvature are significant and complex. Bradshaw [22] has provided a comprehensive review of the longitudinal streamline curvature on turbulent flow. The review shows that the
streamline curvature imposes an extra rate of strain, which is often an order
of magnitude more important than the explicit effects of the extra terms that
appear in the equations of motion. This suggests that the first order solutions
obtained by Karman integral equations would be inadequate and second
order terms should be retained.

To capture the curvature effects properly, we have to appreciate the distinction between the curvature on a body and the curvature of the stream surface in a flow around the body. While the curvature of the physical body may
be accurately defined, the curvature of the stream surface may be dependent
on a number of varying factors, such as the angle of incidence of the incom-
ing flow impinging on the body, or the state of the body, i.e., whether it is
stationary or rotating.

It has been a normal practice to liken the longitudinal and transverse curvature effects to boundary layer flows and generate relevant equations following approaches similar to boundary layer approximations.

Using the above approach, two orthogonal directions may be found at any
point on a three-dimensional surface in which the curvatures are maximum
and minimum. The curvature effects are then captured using a "triply-
orthogonal" coordinate system formed by three families of surfaces, with
coordinates, x_i ($i = 1, 2, 3$). In this coordinate system, an elemental length ds
in space is given by:

$$(ds)^2 = (h_1 dx_1)^2 + (h_2 dx_2)^2 + (h_3 dx_3)^2 \qquad (2.27)$$

where h_i are the metric coefficients corresponding to the coordinate x_i.

The curvature of the surfaces is essentially the rate at which the orthogonal surfaces converge or diverge and can be given by, K_{ij}, such that:

$$K_{ij} = \frac{dh_i}{dx_j}, \text{ for } i \neq j \tag{2.28}$$

The boundary layer on an arbitrary three-dimensional body can be studied using a body-fitted coordinate system where the body surface fits or is represented by one of the surfaces of the triply-orthogonal coordinate system.

Howarth [23] has shown that a coordinate system that is orthogonal everywhere can be constructed around the body with the curves $x_1 = $ constant and $x_3 = $ constant coinciding with the lines of the principal curvatures of the body.

Patel and Sotiropoulos [24] have found such a generalized orthogonal coordinate system very restrictive in boundary layer analysis. For thin boundary layers, it was found to be sufficient to choose a parallel coordinate system in which the parametric curves x_1 and x_3 form an orthogonal net on the surface and the curve x_2 perpendicular to the surface. By considering the relative of orders of magnitude (1 for significant and ε for less significant) of the basic flow parameters, such as velocity and pressure as well as the metric coefficients and curvatures, they were able to obtain the necessary boundary layer equations by reducing the Navier–Stokes for laminar flow and applying the Reynolds averaging approach to turbulent flows.

2.1.6.1 First Order Boundary Layer Equations

Without going into the details of the derivations, we will describe here the final form of the Reynolds averaged turbulent boundary layer equations.

2.1.6.1.1 Continuity Equation

$$\frac{1}{h_1}\frac{\partial U_1}{\partial x_1} + \frac{\partial U_2}{\partial x_2} + \frac{1}{h_3}\frac{\partial U_2}{\partial x_3} + \left(K_{31}U_1 - K_{13}U_3\right) = 0 \tag{2.29}$$

2.1.6.1.2 Momentum Equation in the x_1 Direction

$$\frac{U_1}{h_1}\frac{\partial U_1}{\partial x_1} + U_2\frac{\partial U_1}{\partial x_2} + \frac{U_3}{h_3}\frac{\partial U_1}{\partial x_3}$$

$$+\left(K_{13}U_1 - K_{31}U_3\right)U_3 \tag{2.30}$$

$$+\frac{1}{h_1}\frac{\partial p}{\partial x_1} + \frac{\partial}{\partial x_2}\left(\overline{u_1u_2}\right) - \frac{1}{Re}\frac{\partial^2 U_1}{\partial x_2^2} = 0$$

2.1.6.1.3 Momentum Equation in the x_2 Direction

$$\frac{\partial}{\partial x_2}\left(p + \overline{u_2^2}\right) = 0 \tag{2.31}$$

2.1.6.1.4 *Momentum Equation in the x_3 Direction*

$$\frac{U_1}{h_1}\frac{\partial U_3}{\partial x_1}+U_2\frac{\partial U_3}{\partial x_2}+\frac{U_3}{h_3}\frac{\partial U_3}{\partial x_3}$$

$$+\left(K_{31}U_3-K_{13}U_1\right)U_1 \tag{2.32}$$

$$+\frac{1}{h_3}\frac{\partial p}{\partial x_3}+\frac{\partial}{\partial x_2}\left(\overline{u_2u_3}\right)-\frac{1}{Re}\frac{\partial^2 U_3}{\partial x_2^2}=0$$

2.1.6.2 Some Comments on the Above Equations

We will now comment on some of the aspects of the boundary layer equations:

- The second metric coefficient has been set to a value of unity, that is, $h_2 = 1$. Under this condition, x_2 becomes the physical distance along the line drawn perpendicular to the local surface.

- The pressure variation across the turbulent boundary layer is of the order of ε, but the gradient is of the order of 1. This implies that the boundary layer approximation is more effective in laminar flows than turbulent flows.

  In both laminar and turbulent boundary layers, however, the derivative of the pressure term is zero, suggesting that the pressure remains constant from a point on the surface along a normal direction.

  For a given velocity distribution at the edge of the boundary layer, $U_{1e}(x_1,x_3)$, the inviscid analysis can be used to obtain the pressure distribution, $p(x_1,x_3)$.

- The terms $\dfrac{\partial U_1}{\partial x_3}$ and $\dfrac{\partial U_3}{\partial x_3}$ represent the components of the mean rate-of-strain tensor.

- The terms $-\overline{u_1u_2}$ and $-\overline{u_2u_3}$ represent the components of the Reynolds stress tensor or turbulent stresses.

- The curvature terms K_{13} and K_{32} are absent in the above formulation of boundary layer equations. This is possible if turbulent fluctuations are small.

  We will consider in the next section the cases when the curvature terms need to be retained.

- Often the molecular and turbulent stresses are expressed together as one total stress term. The total stresses denoted by symbols τ_1 and τ_3 in the x_1 and x_3 directions combined this way are given by:

$$\tau_1 = \frac{1}{Re}\frac{\partial U_1}{\partial x_2}-\overline{u_1u_2} \tag{2.33}$$

$$\tau_3 = \frac{1}{Re}\frac{\partial U_3}{\partial x_2} - \overline{u_2 u_3} \tag{2.34}$$

The equations (2.28) and (2.30) can then be re-written as:

$$\frac{U_1}{h_1}\frac{\partial \overline{u_1^2}}{\partial x_1} + \frac{U_2}{h_2}\frac{\partial \overline{u_1^2}}{\partial x_2} + \frac{U_3}{h_3}\frac{\partial \overline{u_1^2}}{\partial x_3}$$

$$+2\left(K_{12}U_1 - K_{21}U_2\right)U_3 \tag{2.35}$$

$$+\frac{1}{h_1}\frac{\partial p}{\partial x_1} - \frac{\partial \tau_1}{\partial x_2} = 0$$

and

$$\frac{U_1}{h_1}\frac{\partial U_3}{\partial x_1} + U_2\frac{\partial U_3}{\partial x_2} + \frac{U_3}{h_3}\frac{\partial U_3}{\partial x_3}$$

$$+\left(K_{31}U_3 - K_{13}U_1\right)U_1 \tag{2.36}$$

$$+\frac{1}{h_3}\frac{\partial p}{\partial x_3} + \frac{\partial \tau_3}{\partial x_2} = 0$$

- At a wall, the turbulent shear stress is zero (no-slip condition), so that:

$$\tau_{w1} = \frac{1}{Re}\left(\frac{\partial U_1}{\partial x_2}\right)_{x_2=0} \tag{2.37}$$

$$\tau_{w3} = \frac{1}{Re}\left(\frac{\partial U_3}{\partial x_2}\right)_{x_2=0} \tag{2.38}$$

- The approximations applied to the Navier–Stokes equations and the turbulent boundary layer equations obtained above are applicable to mean flow conditions.

2.1.6.3 Second Order Boundary Layer Equations

We have already noted that the curvature terms K_{13} and K_{32} can be neglected in the boundary layer equations if turbulent fluctuations are small. This is often not the case.

During the Coanda effect, the entrainment process may produce higher turbulent fluctuations. In such circumstances, these terms need to be retained to reflect the corresponding curvature effects in the boundary layer equations.

Patel and Sotiropoulos [24] have examined this issue and developed the second order boundary layer equations by incorporating the effects of the curvature terms of K_{13} and K_{32}. The resulting equations are given below:

2.1.6.3.1 Continuity Equation

$$\frac{1}{h_1}\frac{\partial U_1}{\partial x_1} + \frac{\partial U_2}{\partial x_2} + \frac{1}{h_3}\frac{\partial U_2}{\partial x_3} + K_{31}U_1 + (K_{32} + K_{12})U_2 + K_{13}U_3 = 0 \qquad (2.39)$$

2.1.6.3.2 Momentum Equation in the x_1 Direction

$$\frac{U_1}{h_1}\frac{\partial U_1}{\partial x_1} + U_2\frac{\partial U_1}{\partial x_2} + \frac{U_3}{h_3}\frac{\partial U_1}{\partial x_3}$$

$$+ K_{12}U_1U_2 + \left(K_{13}U_1 - K_{31}U_3\right)U_3$$

$$+ \frac{1}{h_1}\frac{\partial}{\partial x_1}\left(p + \overline{u_2^2}\right) + \frac{\partial}{\partial x_2}\left(\overline{u_1 u_2}\right) + \frac{1}{h_3}\frac{\partial}{\partial x_3}\left(\overline{u_1 u_3}\right)$$

$$+ \left(2K_{12} + K_{32}\right)\overline{u_1 u_2} + 2K_{12}\overline{u_1 u_3} + K_{31}\left(\overline{u_1^2} - \overline{u_3^2}\right) \qquad (2.40)$$

$$- \frac{1}{Re}\left[\frac{\partial^2 U_1}{\partial x_2^2} + (K_{12} + K_{32})\frac{\partial U_1}{\partial x_2}\right]$$

$$= 0$$

2.1.6.3.3 Momentum Equation in the x_2 Direction

$$\frac{\partial}{\partial x_2}\left(p + \overline{u_2^2}\right) - K_{32}U_3^2 - K_{12}U_1^2 = 0 \qquad (2.41)$$

2.1.6.3.4 Momentum Equation in the x_3 Direction

$$\frac{U_1}{h_1}\frac{\partial U_3}{\partial x_1} + U_2\frac{\partial U_3}{\partial x_2} + \frac{U_3}{h_3}\frac{\partial U_3}{\partial x_3}$$

$$+ \left(K_{31}U_3 - K_{13}U_1\right)U_1 + K_{32}U_3U_2$$

$$+ \frac{1}{h_1}\frac{\partial}{\partial x_1}\left(\overline{u_1 u_3}\right) + \frac{\partial}{\partial x_2}\left(\overline{u_2 u_3}\right) + \frac{1}{h_3}\frac{\partial}{\partial x_3}\left(p + \overline{u_3^2}\right)$$

$$+ 2K_{31}\overline{u_1 u_3} + \left(2K_{32} + K_{12}\right)\overline{u_2 u_3} + K_{13}\left(\overline{u_3^2} - \overline{u_1^2}\right) \qquad (2.42)$$

$$- \frac{1}{Re}\left[\frac{\partial^2 U_3}{\partial x_2^2} + (K_{12} + K_{32})\frac{\partial U_3}{\partial x_2}\right]$$

$$= 0$$

2.1.6.4 Reduction of Second Order Boundary Layer Equations for Two Dimensions

2.1.6.4.1 Continuity Equation

$$\frac{1}{h_1}\frac{\partial U_1}{\partial x_1} + \frac{\partial U_2}{\partial x_2} + K_{13}U_3 = 0 \qquad (2.43)$$

2.1.6.4.2 Momentum Equation in the x_1 Direction

$$\frac{U_1}{h_1}\frac{\partial U_1}{\partial x_1} + U_2\frac{\partial U_1}{\partial x_2}$$

$$+K_{12}U_1U_2 + \frac{1}{h_1}\frac{\partial}{\partial x_1}\left(p + \overline{u_2^2}\right)$$

$$+\frac{\partial}{\partial x_2}\left(\overline{u_1 u_2}\right) + \frac{1}{h_3}\frac{\partial}{\partial x_3}\left(\overline{u_1 u_3}\right) \qquad (2.44)$$

$$+2K_{12}\overline{u_1 u_2}$$

$$-\frac{1}{Re}\frac{\partial}{\partial x_2}\left[\frac{\partial U_1}{\partial x_2} + K_{12}U_1\right] = 0$$

2.1.6.4.3 Momentum Equation in the x_2 Direction

$$\frac{\partial}{\partial x_2}\left(p + \overline{u_2^2}\right) - K_{12}U_1^2 = 0 \qquad (2.45)$$

2.1.7 Special Mathematical Models for Blowing

Here we present two models to study the vortex wake characteristics from the effects of blowing:

- Enhanced viscous diffusion from blowing
- Point vortex model

2.1.7.1 Viscous Diffusion

We will start with the vorticity equation, which is equally as significant as a momentum equation. If we consider a Newtonian fluid of fixed density and viscosity, then the vorticity equation for potential body forces can be written as [11]:

$$\frac{D\vec{\omega}}{Dt} = \left(\vec{\omega}.\nabla\right)\vec{u} + v\nabla^2\vec{\omega} \qquad (2.46)$$

where,

$\dfrac{D\vec{\omega}}{Dt}$ is the rate of change of particle vorticity

$(\vec{\omega}.\nabla)\vec{u}$ is the rate of deforming vortex lines

$\nu\nabla^2\vec{\omega}$ is the net rate of viscous diffusion of $\vec{\omega}$

Expanding Equation (2.46),

$$\frac{D\vec{\omega}}{Dt} + (\vec{u}.\nabla)\vec{\omega} = (\vec{\omega}.\nabla)\,\vec{u}) + \nu\nabla^2\vec{\omega} \tag{2.47}$$

If the second and third terms of this equation are ignored, then the equation represents convection diffusion or vorticity diffusion through viscosity. The reduced form of the equation, generally known as the Poisson equation, becomes:

$$\frac{D\vec{\omega}}{Dt} = \nu\nabla^2\vec{\omega} \tag{2.48}$$

Lamb [25] was probably the first to use the above equation to model viscous laminar decay of a two-dimensional line vortex. He obtained a solution of the form:

$$\zeta(r,t) = \frac{\Gamma_1}{4\pi\nu t}\exp\left(\frac{-r^2}{4\nu t}\right) \tag{2.49}$$

The kinematic viscosity in the above expression appears in the denominator of the exponential term, suggesting that the decay of the vortex in a laminar viscous diffusion would be a slow process. Following the discussion of Squire [26], it appears that a better representation of the vortex decay would be to replace the kinematic viscosity, ν, by an equivalent viscosity, ν_T.

The resulting tangential velocity distribution can be written as:

$$U_t(r,t) = \frac{\Gamma_1}{2\pi r}\exp\left(1 - \frac{r^2}{4\nu_T t}\right) \tag{2.50}$$

2.1.7.2 Point Vortex Model

Westwater [27] was probably the first to attempt the point vortex method to model and investigate the effect of discrete jet blowing on the characteristics of vortex–wake rollup. Despite being a very idealized representation of the characteristics, the method has been found to provide useful pointers regarding them.

The method assumes an inviscid flow where the time-dependent rollup of a vortex sheet is represented by an equivalent array of two-dimensional line vortices. A pair of counter rotating vortices are introduced to model the effect of discrete jet blowing.

Downstream of the wing, the line vortices are assumed to be represented by point vortices. The strength of each vortex is dependent on the bound circulation distribution.

In applying this method, the vortex sheet is divided into a number of discrete points, say n, where each vortex has a strength of Γ_j. The bound circulation is discretized over each subdivision to determine the vortex strength of each vortex.

The Biot-Savart law is used to determine the velocities induced on each other at the point vortices. The resulting expressions obtained are:

$$\frac{dy_i}{dt} = -\sum_{j\neq 1}^{n} \frac{\Gamma_j}{2\pi} \frac{z_i - z_j}{\left(y_i - y_j\right)^2 + \left(z_i - z_j\right)^2} \tag{2.51}$$

$$\frac{dz_i}{dt} = -\sum_{j\neq 1}^{n} \frac{\Gamma_j}{2\pi} \frac{y_i - y_j}{\left(y_i - y_j\right)^2 + \left(z_i - z_j\right)^2} \tag{2.52}$$

Integrating Equations (2.51) and (2.52) with respect to time, the movement of the point vortices are obtained. The process is repeated for a new vortex position.

Westwater [27] had assumed that the strength of the vortex at each subdivision was equal in magnitude. He produced results that showed the formation of smooth spirals from vortex rollup. These results were difficult to reproduce in subsequent attempts by other works [28, 29]. It was concluded that that the large time-steps and integration method adopted by Westwater may have contributed to smooth but inaccurate results.

Another problem with the method arose from the chaotic motion observed at the tip region. The spiral produced was, however, nearly asymmetric in nature, leading Moore [28] to suggest that it would be better to combine the finite number of vortices into a single vortex center rather than have them scattered on the spiral. This approach helped to suppress the chaotic motion produced near the tip. The method then produced smooth spirals and detailed rollup of vortices.

2.2 Physical Experimentation

In this section, we will introduce some important aspects of physical experimentation that would be useful in any Coanda effect investigation. We will conduct the discussion under the following headings:

- Facilities for controlled experiment
- Flow diagnostic techniques
- Reduction of data and analysis

2.2.1 Facilities for Controlled Experiment

A common experimental facility is a wind tunnel. This facility offers the opportunity to conduct fluid mechanical investigations under controlled environment. The need for such experiments arises because they offer greater control over the effects of the features under investigation.

The basis for modern wind tunnels can be traced back to the Wright brothers who conducted controlled experiments to study the behavior of lifting bodies and their flight mechanism. Their efforts eventually paved the way for the first successful manned flight in human history.

A main feature of any wind tunnel is its "test section" where the flow is maintained as uniform and one-dimensional. When a test model is placed in this section, the flow behavior is altered. By comparing the behavior of the altered flow with the original flow, the forces and their moments acting on the body are assessed and their characteristics determined.

We have to decide first the objective of the experiment and the flow parameters that we want to evaluate. The type, size, and capabilities of the testing facility become the next set of considerations as they dictate the configuration details of the test models that need to be constructed and the instrumentation required for testing.

Various types of wind tunnels [30–32] such as subsonic, transonic, supersonic, or hypersonic flows have been designed over the last hundred years and have reached, somewhat, a state of maturity. These facilities, however, are expensive to construct, operate, and maintain. Thus, particular attention to cost is also essential.

Due to cost considerations, constructed test models must be small but representative of the real case, and adequate plans must be made to conduct a minimum number of tests without compromising the quality of the data. Careful pre-planning and greater appreciation of issues involved are, therefore, necessary.

Dimensional analysis is often a procedure that may be used in any experimental program because such analysis may lead to a fewer number of tests by combining several variables to be grouped together and expressed as functions of non-dimensional numbers.

In many complex flow studies, a qualitative assessment of the nature of the flow can assist in the proper formulation of the quantitative test program that may follow. The qualitative assessment may take the form of a flow visualization investigation where smoke, tufts, surface oil, or a laser light sheet are used [33–36]. Figure 2.1 (a) and Figure 2.1 (b) show sketches of tufts depicting attached flow and separated flow, respectively, on a surface.

In quantitative investigations, the choice of flow diagnostic technique and instrument become important considerations. If we take the case of a body immersed in a stream, lift and drag forces would be the most important information we would seek. We can, for instance, use a force balance to obtain these forces. This will provide the magnitude of total lift or drag

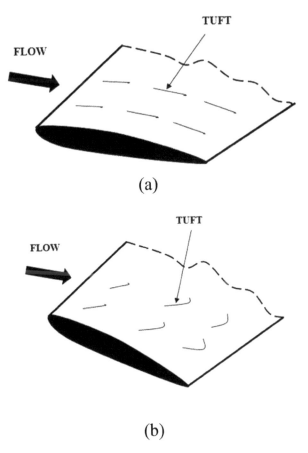

FIGURE 2.1
Flow visualization using tuft, showing the (a) flow remaining attached to a surface and (b) the flow separating from a surface.

forces acting on a body. However, if we want to obtain detailed information of the flow behavior that is related to these forces, such as regions of suction, flow deceleration, or breakdown, we may have to resort to other techniques, such as obtaining information of the pressure distribution or temperature distribution on or around the surface of the body. In such cases, we may install pressure tapping points or hot films on the body itself. Or we may conduct pressure or velocity measurements very near to the surface or away from it to gather the necessary information.

2.2.2 Flow Diagnostic Techniques

In this section, we will describe three standard flow measurement techniques, namely, pressure based, hotwire anemometry, and laser anemometry.

These are point-flow measurement techniques. The first two are intrusive while the third is a non-intrusive technique in nature.

2.2.3 Pressure-Based Measurement Technique

The origin of pressure-based techniques goes back nearly three hundred years. In 1732, Henry Pitot discovered [37] a direct connection between pressure and velocity, which was modified nearly a century later by Darcy [38], which subsequently formed the basis of pressure and velocity measurement at a point.

2.2.3.1 One-Dimensional Velocity Measurement

The early pressure probe that emerged as a consequence of the Pitot discovery was a single hole probe, the Pitot probe, capable of measuring the total or stagnation pressure of a flow. The probe, however, had to be aligned in such a way that its hole faced the oncoming flow directly. With the addition of another hole or a static port, the probe evolved into a two-hole or a Pitot-static probe. Since this probe could now be used to measure both the static and total pressure at a point, it provided a simple method to deduce the velocity at that point. Figure 2.2 shows a schematic of how the original Pitot tube with the introduction of a static port has become a Pitot-static tube [39].

According to the law of conservation of energy, the total energy at any point in a fluid stream remains constant. If this energy is expressed in terms of pressure (energy per unit volume), we can express ignoring losses, P_{total}, the total pressure, as being made up of P_{static}, the static pressure, and $P_{dynamic}$, the dynamic pressure. Since the magnitude of the dynamic pressure is given by $\frac{1}{2}\rho U_\infty^2$, we can write:

$$P_{total} = P_{static} + \frac{1}{2}\rho U_\infty^2$$

rearranging,

$$U_\infty = \sqrt{\frac{2\left(P_{total} - P_{static}\right)}{\rho}}$$

The fact that the Pitot technique has survived to this day is a testament to the reliability and robustness of the technique. Other flow-measuring techniques are often calibrated against measurements obtained from this technique. Even today, in aviation, a Pitot-static system is used to provide the airspeed, Mach number, or altitude of an aircraft in flight. The total and static pressures, however, are obtained separately using a total head Pitot tube aligned in the flight direction while the static

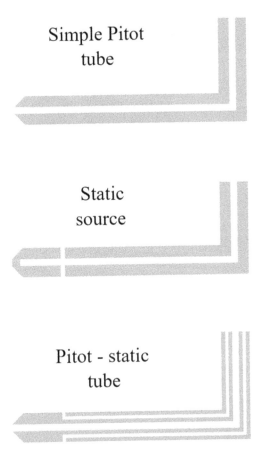

FIGURE 2.2
A Pitot-static tube. (After: Chaos386, 2007 [39].)

pressure is obtained from an additional static port located on the aircraft body. Figure 2.3 shows a schematic of the Pitot-static system used in aviation that consists of a Pitot tube, a static port, and Pitot-static instruments [40].

Although there have been modifications and improvements to the Pitot probes, the underlying concept has remained the same. We will now discuss some of these improvements and developments.

In 1910, Prandtl made a significant modification to the shape of the Pitot-static probe and the location of the static pressure port based on his boundary layer theory. This made the measurement of the total and static pressures more reliable, and this modified version of the Pitot-static probe is widely used [41] today in fluid mechanical measurements.

The limitation of the Pitot-static probe is that it is one-dimensional, i.e., capable of measuring one component of the velocity or when aligned to the

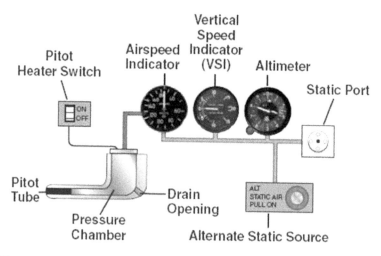

FIGURE 2.3
Diagram of a pitot-static system used in aviation. (After: [40].)

direction of the flow. A Pitot-static probe is well suited to measure velocity of the wind tunnel at the test section because the flow in it, by design, is one-dimensional and uniform. In practical application, the direction of the flow is often not known; the probe has to be aligned with the flow direction by trial and error.

2.2.3.2 Two-Dimensional Velocity Measurement

Yaw or three-hole probes have been developed to measure two-dimensional flows. The limitation of this probe is that it has to be nulled, that is, two of its side holes located at equal angle and distance from the central hole have to be made equal by trial and error rotation in the two-dimensional plane of measurement. This helps to determine the angle and total velocity of the flow and from them its two-dimensional velocity components. Other methods using three and four holes have also been attempted for two- and three-dimensional velocity component measurements, mostly with nulling requirement.

2.2.3.3 Three-Dimensional Velocity Measurement

Five-hole and seven-hole pressure probes were the natural extensions of pressure-based methods to obtain velocity components in three-dimensions. These probes can be considered more user-friendly because they can be used in non-nulling mode in a flow where the direction of the flow is not known.

The drawback of the non-nulling method, however, is its dependence on calibration curves that have to be generated prior to an actual experiment. From these calibration curves, the total pressure, static pressures, and flow angle can be obtained, and thence, the velocity components in the three directions. A sample of the three calibration curves [42] is shown in Figures 2.4, where P_1, P_2, P_3, P_4 and P_5 denote the ports of the five holes on a probe head such that P_1 is located at the center, P_2 and P_3 on the yaw plane, and P_4 and P_5 on the pitch plane.

Three calibration curves are produced by placing the probe and gathering data at some pre-determined values of pitch α and yaw angles β. The pressure readings from each of the ports are expressed in non-dimensional forms of four parameters, namely:

$$C_{p\text{Yaw}}, C_{p\text{Pitch}}, C_{\text{Static}}, \text{ and } C_{p\text{Total}}$$

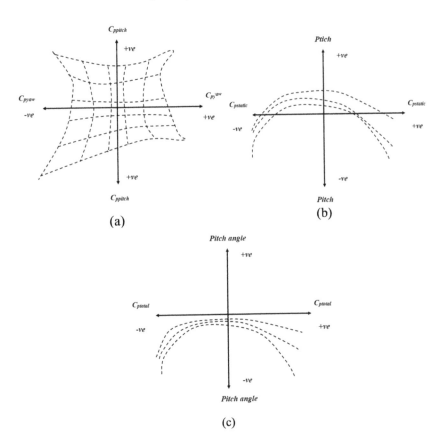

FIGURE 2.4
Calibration charts of a five-hole probe. (a) C_{pyaw} versus C_{ppitch}; (b) C_{pstatic} versus α; (c) C_{ptotal} versus α. (After: [42].)

where,

$$C_{p\text{Yaw}} = \frac{\left(P_2 - P_3\right)}{\left(P_1 - \bar{P}\right)}$$

$$C_{p\text{Pitch}} = \frac{\left(P_4 - P_5\right)}{\left(P_1 - \bar{P}\right)}$$

$$C_{p\text{Static}} = \frac{\left(\bar{P} - P_{\text{Static}}\right)}{\left(P_1 - \bar{P}\right)}$$

$$C_{p\text{Total}} = \frac{\left(P_1 - P_{\text{Total}}\right)}{\left(P_1 - \bar{P}\right)}$$

and,

$$\bar{P} = \frac{\left(P_1 + P_2 + P_3 + P_4\right)}{4} \tag{2.53}$$

It is worth noting that the calibration curves are not symmetric about the lines of zero angles of pitch and yaw. Also in these curves, as the values of pitch and yaw angles become larger, the values of $C_{p\text{yaw}}$, $C_{p\text{pitch}}$, $C_{p\text{static}}$, and $C_{p\text{total}}$ also become very large. Beyond the values of pitch and yaw angles of ±10°, the errors in the deduced pressure and velocity components start becoming very large, and beyond ±30° highly unreliable.

Pisasale and Ahmed [43] found that the cause of the above behavior of the calibration curves was due to the singularity caused in the denominator when $\left(P_1 - \bar{P}\right) \to 0$. They therefore conducted a theoretical calibration [44] on a five-hole probe with a spherical probe head, shown in Figure 2.5 and produced trends similar to those obtained in practice (Figure 2.4). The results are shown in Figure 2.6.

From their theoretical studies, Pisasale and Ahmed [43] were able to confirm the existence of singularity in the conventional calibration procedure of defining the denominator used in the determination of $C_{p\text{yaw}}$, $C_{p\text{pitch}}$, $C_{p\text{static}}$, and $C_{p\text{total}}$. The authors found that the singularity was found to occur at ±54.7° by plotting a curve, $C_{p\text{den}}$, against α, as shown in Figure 2.7, where,

$$C_{p\text{den}} = \frac{P_1 - \bar{P}}{q} \text{ and } q = \frac{1}{2}\rho_\infty U_\infty^2.$$

Based on the above work, Pisasale and Ahmed [44] proposed a new procedure to overcome the singularity problem in calibration of a five-hole probe and gave a detailed step-by-step procedure of how to use this technique. Here we will highlight the basis of their procedure.

In the proposed procedure, the denominator was treated as a single unknown and its magnitude was determined by establishing a functional

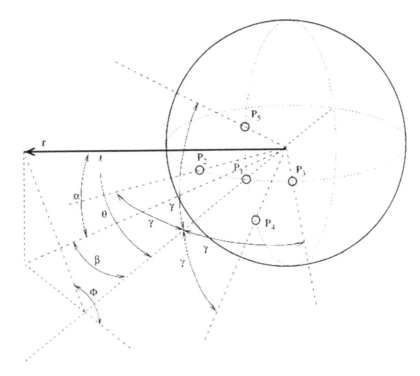

FIGURE 2.5
A schematic showing locations of the holes of a five-hole pressure probe. (After: Pisasale and Ahmed, 2002 [43].)

relationship between the denominator and the pressures recorded in a real flow. For the functional relationship, the following conditions were imposed:

- Any combination of pressures used to form a parameter must be independent of any reference pressure or Reynolds number. This will make the relationship between the denominator and the measured pressures valid in all unknown flows, and will allow the probe to be used regardless of the particular flow conditions. Consequently, pressure parameters will be non-dimensional.
- The relationship must be true for any combination of pitch angle and yaw angle. That is to say that the relationship between parameters are not functions of α or β, although the parameters themselves may be a function of pitch and yaw angles. This is necessary as neither the pitch nor yaw angles are known before the calibration coefficients are calculated.
- Individual parameters must avoid singularity wherever possible. This will allow the procedure to be used over the greatest possible range of flow angles, without significant loss of accuracy.

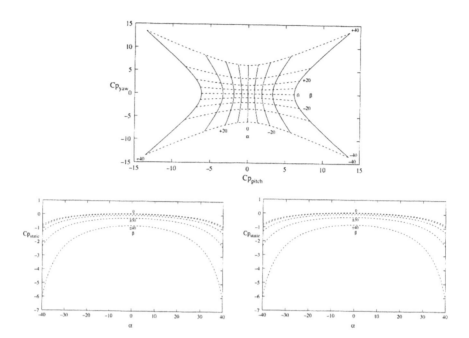

FIGURE 2.6
The three theoretical calibration curves for a five-hole probe. (After: Pisasale and Ahmed, 2002 [43].)

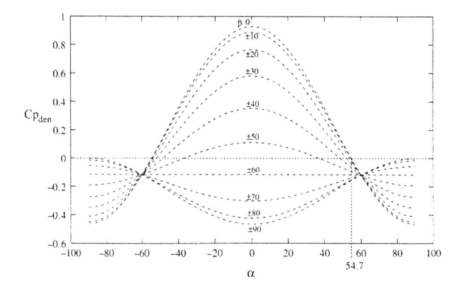

FIGURE 2.7
Demonstration of singularity in the denominator in calibration. (After: Pisasale and Ahmed, 2002 [43].)

The following two parameters were found to meet most of the above requirements [44]:

$$N_y = \frac{P_1 - \bar{P}}{\text{DEN}},$$

where, $\text{DEN} = P_1 - P_{\text{static}} + Aq$,

$$N_x = \frac{P_1 - \bar{P}}{\sqrt{(P_2 - P_3)^2 + (P_4 - P_5)^2}}$$

The functional relationship between the parameters, N_y and N_x, using data from a range of flow conditions were plotted that confirmed that these parameters were independent of the Reynolds number, pitch angle, and yaw angle. The plot is shown in Figure 2.8.

In Figure 2.8, a singularity in N_x can be observed near zero pitch and yaw angles. The N_y values are, however, nearly constant in this region and standard calibration methods [44] can be used instead.

The other point worth noting in Figure 2.8 is that the DEN terms become indeterminate when $N_y = 0$. An alternative, functional relationship between N_{y*} and N_x was suggested to resolve this issue, to obtain the value of DEN with,

$$N_{y*} = \frac{\text{DEN}}{\sqrt{(P_2 - P_3)^2 + (P_4 - P_5)^2}}$$

This new relationship is given in Figure 2.9. Values for DEN can be obtained using either Figure 2.8 or Figure 2.9, but there is more scatter in the data in Figure 2.9 than in Figure 2.8.

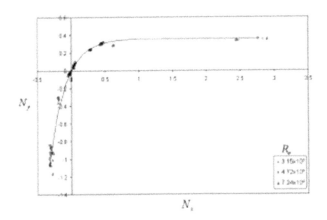

FIGURE 2.8
Functional relationship curve for N_y and N_x. (After: Pisasale and Ahmed, 2002 [44].)

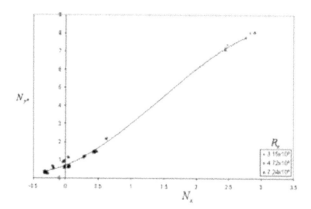

FIGURE 2.9

Functional relationship curve for N_{y*} and N_x. (After: Pisasale and Ahmed, 2002 [44].)

It is generally assumed that the multi-hole pressure probe head would be placed in a stream of flow, even when high flow angularity is encountered, with most of the pressure ports exposed to the oncoming flow directly. In cases where the probe head is erroneously placed, such as in a reversed flow, then the readings recorded in all of the pressure ports may become the same. The objective would, therefore, be to correctly orient the probe head in the direction of flow.

One possible step would be to rotate the probe head through 180° and compare the measured pressures at this position with those obtained in the previous position and make a judgement about the correct direction based on the calibration of the probe. The flow may also be highly unsteady and the use of different measurement instruments may be more effective. These issues have been discussed in more detail in Reference [45, 46].

2.2.3.4 Skin Friction Measurement

In a pressure-driven flow field, the introduction of an intrusive technique, such as a pressure probe, may produce error in the measurement due to flow interactions between the stream lines and the probes. Such effects, however, are not severe, and in shear driven three-dimensional boundary layers or in cases where the curvature is not very strong, as the studies of Vagt and Fernholz [47] suggest, the probe effect can be neglected.

With the appropriate choice of probe geometry along with a high-quality pressure transducer or micro manometer, it has also been demonstrated [47, 48] that it is possible to measure flow angles in turbulent boundary layers quite accurately, to within ±0.1°.

Various direct and indirect methods are available for skin friction measurements [49–53]. Compared to the indirect methods, direct methods are difficult to implement. Indirect methods use devices such as a Preston tube

[54], Stanton tube [55], sub-layer fence [56], and so forth. These methods use analytical correlations to relate the total or static pressure or heat transfer data to evaluate the skin friction.

The indirect skin friction methods require the similarity law or the logarithmic law of the wall to be valid in the boundary layer near the wall. If the flow is one-dimensional or if the flow angles are restricted below ±15° [57], then the Preston method [56] appears to be quite effective in measuring skin friction. Several attempts [58, 59] to extend the Preston method, however, to flows with higher flow angularity have not been very successful.

Lien and Ahmed [60] found that it was possible to apply the five-hole pressure technique in two-dimensional skin friction measurement. They thereafter attempted to measure shear stress angles to determine the skin friction in three-dimensional turbulent boundary layers [61].

In three-dimensional flows, boundary layers may be skewed and a knowledge of their velocity profiles can help determine the shear stress angle near the wall as the ratio of velocity components, w and u, in the crosswise and streamwise directions.

Figure 2.10 shows velocity profiles in the streamwise and cross flow direction, where, γ_w is the shear stress angle, which can be given by:

$$\gamma_w = \tan^{-1}\left(\frac{w}{u}\right)$$

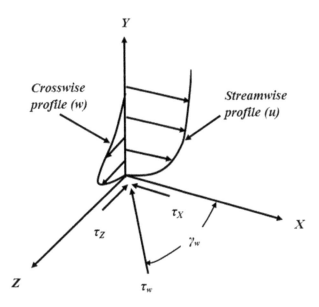

FIGURE 2.10
Schematic of velocity profiles in a three-dimensional turbulent boundary layer.

When a five-hole pressure probe is used (with reference to Figure 2.5) and if the total velocity is U_∞, the velocity components, u and w can be obtained as:

$$u = U_\infty \cos\alpha \sin\beta \qquad\qquad (2.54)$$

$$w = U_\infty \sin\alpha \qquad\qquad (2.55)$$

From Equations (2.54) and (2.55), the yaw angle measured by the five-hole pressure probe is then the shear stress angle and can be expressed as:

$$\gamma_w = \tan^{-1}\left(\tan\alpha\cos\beta\right)$$

The wall effect on the readings of a five-hole pressure probe is shown in Figure 2.11 (a) and Figure 2.11 (b). When there is a gap between the wall and

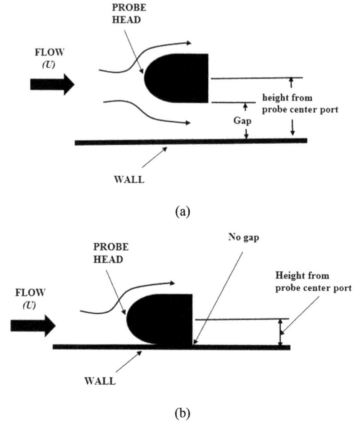

FIGURE 2.11

A schematic of the wall effect on pressure readings on a five-hole pressure probe reading: (a) gap exists between the wall and the probe; (b) no gap exists between the wall and the probe.

probe, as shown in Figure 2.11 (a), flow can move around the probe head. However, when there is no gap between the wall and probe, as shown in Figure 2.11 (b), the flow movement in the streamwise direction is affected, which in turn affects the reading from each port of the multi-hole probe.

Based on one-dimensional pipe flow measurement, Lien and Ahmed [61] determined that when a five-hole probe was placed in line with the flow, the maximum departure of pitch and yaw angles from the zero value was around 4° and 2°, respectively. This implied a greater impact of wall on pitch angle readings than the yaw angle readings of the five-hole probe, which translated to ±0.3% error in the raw pressure readings, which was on par with the Preston tube readings. Since the five-hole probe application is also based on the similarity principle, the method would also be effective for probes with different geometries and sizes provided they were small enough to be submerged in the boundary layer and had reliable calibration coefficients.

2.2.3.5 Fluctuation Considerations

The pressure-based measurements mentioned above are not effective in providing instantaneous information of a flow. In a turbulent flow, the fluctuating pressures at the sensing points are generally damped out and the differential pressures recorded will be of steady values with the effect of turbulence imbedded. Some form of corrections may be attempted from the following considerations:

Let us re-consider Bernoulli's equation in the form:

$$P_{total} = P_{static} + \frac{1}{2}\rho U_\infty^2$$

By decomposing each of the flow parameters in terms of their mean and fluctuating components,

$$P_{total} = \overline{P}_{total} + p'_{total}$$

$$P_{static} = \overline{P}_{static} + p'_{static}$$

$$u = \overline{U}_\infty + u' + v' + w'$$

$$\rho = \overline{\rho} + \rho'$$

We get:

$$\overline{P}_{total} + p'_{total} = \overline{P}_{static} + p'_{static} + \frac{1}{2}\left(\overline{\rho} + \rho'\right)\left(\overline{U}_\infty + u' + v' + w'\right)^2$$

Applying Reynolds rule of averaging and re-arranging,

$$\overline{P}_{total} - \overline{P}_{static} = \frac{1}{2}\left(\overline{\rho}\right)\overline{U}_\infty^2 + \left(A_1\overline{u'^2} + A_2\overline{v'^2} + A_3\overline{w'^2}\right)$$

where the term $\left(A_1 \overline{u'^2} + A_2 \overline{v'^2} + A_3 \overline{w'^2} \right)$ represents the effect of turbulence in the measurements, and the coefficients A_1, A_2, and A_3 may change depending on the turbulent nature of the flow, and are difficult to obtain.

Gatto et al. [62–64] have shown that it is possible to determine both the mean and fluctuating pressures on a surface over which a fluid flows using plastic tubes which connect the surface tapping points to a remote pressure transducer provided the transfer functions that relate the fluctuating pressures at the opposite ends of the tubes are known.

Instantaneous velocity information in a turbulent flow are generally obtained using hot-wire techniques [65, 66] which we will discuss next.

2.2.4 Hot-Wire Anemometer

The hot-wire anemometer, throughout the last hundred years, has proven itself as one of the most effective flow measurement devices that is widely used in laminar, transitional, and turbulent flow investigations. Its ability to determine temporally fluctuating velocity components has made it an indispensable tool in most fluid mechanical investigations.

The hot wire anemometer is a thermal anemometer that depends on convective heat propagation into its surrounding fluid. It consists of a sensor in the form of a very fine wire that is only few microns in diameter and a set of electronic equipment that converts the information of the signal into an electrical signal. Because of the fineness of the hot-wire, it is very delicate and must be handled with care. Many laboratories for this reason are equipped with hot-wire repairing accessories. Industrial versions of a hot wire anemometer are also available where the wire is encased, making it more robust, but this comes at the expense of being larger and providing lower measurement resolution.

The hot wire measures the normal component of the velocity of an oncoming flow. Depending on the number of wires, the hot-wire probes are called single, dual (or X), or triple-wire, and are generally used to obtain a one-, two-, or three-dimensional velocity component of a flow at any instant of time. Often a single hot-wire probe can be rotated and measurements obtained at two locations. The information can be combined to give two velocity data points if the assumption that the flow behavior at the two moments of time of measurement has not changed. Using the same assumption, a slant hot wire can be used to obtain two- or three-dimensional velocity information through rotation at different positions. Often a four-wire probe is used, where the fourth wire is used for additional data for validation of the data.

Figure 2.12 shows a schematic of the various hot-wire probe configurations.

2.2.4.1 *Principle of Operation*

The principle of operation of a hot-wire anemometer can be explained as follows. An electric current passes to heat the hot wire above the ambient

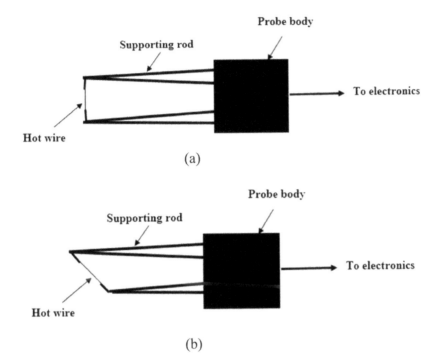

FIGURE 2.12
Examples of hot-wire probes: (a) single normal wire and (b) single slant wire.

temperature when it is placed in fluid stream. The heat is transferred from the wire to the fluid flowing over it. This results in the cooling of the wire, and the thermal balance of the wire changes, which in turn is reflected in changes in the electrical resistance of the wire. The electronic equipment then measures these changes from which the speed of the fluid is extrapolated.

The hot-wire anemometry technique is an indirect method, meaning velocity information cannot be gathered directly, but are obtained in terms of voltage. This necessitates some other method to establish the functional relationship between voltage and velocity by calibrating a given hot-wire anemometer against known velocities.

There are various modes of implementing hot-wire devices. The constant current anemometer (CCA) and the constant temperature anemometer (CTA) are the two most common [67–70], with the CTA being the most commonly used one. Depending on the mode, the circuit within the electronic equipment is used to maintain the current or voltage constant to achieve thermal balance. The performance of the electronic circuitry, therefore, plays an important role in the proper functioning of the hot anemometry system [71]. A classical resistive Wheatstone bridge circuitry used in a constant temperature anemometer is shown in Figure 2.13.

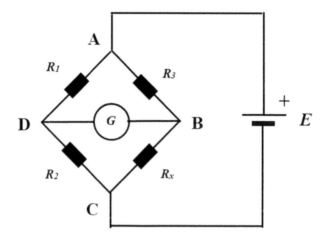

FIGURE 2.13
Schematic of the classical Wheatstone bridge circuitry.

2.2.4.2 Calibration Methods

Various hot-wire calibration methods have been developed with various degrees of complexities and success [72, 73]. In general, differing calibration methods are considered based on whether the velocity range is above or below 1.5 m/s.

For velocities above approximately 1.5 m/s, the hot wire is calibrated against data obtained by placing a Pitot-static probe in a wind tunnel test section or in a uniform steady nozzle flow [74–76]. Figure 2.14 shows a typical setup for this velocity range calibration using calibration equipment available commercially (such as from DANTEC), where the nozzle flow of a

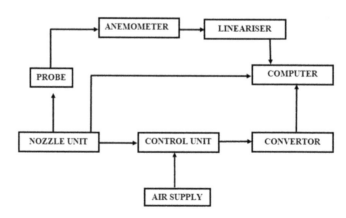

FIGURE 2.14
Block diagram of a hot-wire calibration setup.

known velocity at a particular pressure is used. The nozzle can be changed to produce different velocities.

However, the use of this conventional method becomes ineffective for velocities below 1.5 m/s, when the pressure or velocity reading from the manometer becomes difficult to read accurately. Consequently, various methods have been developed including the laminar pipe flow method [77–79], rotating disc method [80], vortex shedding method [77, 81], swing arm apparatus [82], pendulum technique [83], towing tanks [73, 84], and so forth.

Once the calibration data for a single wire is obtained, it is then fitted to the widely used King's law [74, 84], as expressed below:

$$E^2 = A + BU^n$$

where

    E is the voltage across the wire
    U is the velocity of calibration flow and
    $A, B,$ and n are the King's constants for the particular hot-wire

Various authors have fitted their data to obtain values for n for different ranges of the curve generated. Tsanis [85], for example, separated his velocity range, of 0.04 m/s to 2.5 m/s, into three parts. To simplify the process, often a value of $n = 0.45$ or 0.5 can be fixed and then the values of the constants A and B determined [66].

Various analytical methods have been developed for data processing of the information obtained under static conditions taken from two sensor probe [86–89] or triple or multiple sensor probes [90]. These are time-con-suming processes requiring a large number of data points from calibration for greater precision. But the probe characteristics may change with time due to many factors, such as the deposition of very minute dust particles on the hot-wire altering the characteristics of the calibration curves. In many situa-tions, calibration before and after the experiment are conducted to check for reliability of the calibration data, requiring even more time. Lecic [91] claims to reduce the time of calibration significantly by proposing a new method of dynamic hot-wire calibration in a quasi-stationary air tunnel jet.

For measurements in a fluctuating freestream flow, the dynamic calibra-tion method developed by Perry and Morrison [92] can be used. The pro-cedure involves shaking the hot-wire probe with a small sinusoidal motion and small frequency range, from 0 to 10 Hz, in a uniform flow of known velocity. Apart from introducing complexity in the experimental task, the frequency range is also considered too small for most turbulent flow studies. Hence many workers still use static calibration data based on the assumption that, because the size of its active element is so small, it will have low ther-mal inertia and, therefore, be capable of faithfully capturing the fluctuating components [93].

Measurements near a wall pose severe problems because the probe and its intrusive nature are similar to what we have seen with pressure probes before, i.e., they alter the flow around the probe and near the wall. Although various corrections to the readings of the hot-wire anemometers have been proposed [74, 94], the inherent difficulties associated with experimentation invariably require additional numerical investigation to complement the near-wall study results.

In turbulence measurement, the concept of "conditional sampling" introduced by Kovasznay et al. [95] can be considered a significant development. In conditional sampling, the statistical averages are performed on data gathered over periods of time for which the flow satisfies some condition chosen by the experimenter, such as an "intermittent average" where data is accumulated only over the periods for which the flow at the measurement point is turbulent.

2.2.5 Laser Anemometry

The word LASER is an acronym for "Light Amplification by Stimulated Emission of Radiation." Today LASER is also written in small letters as "laser."

The helium–neon (He–Ne) laser was developed in 1962 at the Bell Telephone Laboratories in the USA. It produced at a wavelength of 632 nm a continuous wave of electromagnetic radiation in the visible spectrum [96]. Two years later, Yeh and Cummins [97] published a paper showing that fluid flow measurement would be possible from the Doppler effect on a He–Ne beam scattered by very small polystyrene small spheres entrained in the fluid. This led to the development, the following year, of the first laser Doppler flowmeter [98] using heterodyne signal processing. Today, the anemometer based on this principle is known as the laser Doppler velocimeter (LDV) or laser Doppler anemometer (LDA). This non-intrusive optical diagnostic technique [99, 100] has found wide-ranging applications and has been used to measure speeds of flow ranging from a few mm/s to 1000 m/s.

Apart from the LDV, the laser two focus velocimeter (L2F) is another technique that has been developed using the laser for velocity measurements (the L2F is also known as two-spot or time of flight laser measurement). Both the LDV or L2F techniques require two exactly similar laser beams in their operation. The action takes place at the waists of the laser beams in both cases. The waist can be likened to the throat of a diverging–converging nozzle where the laser light has the highest intensity. Figure 2.15 shows the waist of a laser beam.

There are three main components of a LDV or L2F system: a laser source, a receiver, and a signal processor. The scattered light signals can be very weak and a photomultiplier tube is used to count the photons and increase their intensity before signal processing. In the case of LDV, the signal processing involves auto-correlation of the data whereas cross-correlation is

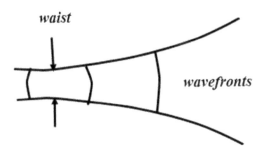

FIGURE 2.15
Waist of a laser beam.

performed between the scattered signals passing through the two waists of a L2F system.

The earlier versions of LDV or L2F equipment had to be placed near the measurement location. The development of single and mono-mode optical fibers and laser diodes and their inclusions has given greater scope to reduce the sizes of the equipment. The transmission of laser light and the receiving of the scattered signal can be performed via these optical fibers, and the experiment can be conducted remotely.

Both the LDV and L2F can be operated in forward scatter, back scatter, or oblique scatter modes depending on the availability of the optical access to receive the scattered signal. The strength of the scattered signal in the forward mode is the highest and can easily be 200 times higher than that at back scatter mode. In most practical applications, however, optical access to receive the scattered signal is very limited, and resorting to back or oblique scatter is the norm.

There is a third optical measurement technique that has been developed and is worth mentioning. This is the particle image velocimetry or the PIV technique. The main difference between the PIV and the LDV or L2F technique is that the PIV is capable of producing two-dimensional or even three-dimensional vector fields, while the other two techniques of LDV and L2F measure the velocity at a point.

All these optical diagnostic techniques require "seeding" or introduction of particles in the flow that are light reflective and faithfully follow the flow without producing any distortion to the flow [101].

2.2.5.1 Laser Doppler Velocimeter

There are various types, such as the reference, two-scatter, or fringe type for which a LDV can be designed and operated. The "fringe" type or mode of operation appears to be the most commonly used and will be discussed here.

In a fringe LDV, two identical laser beams are made to intersect to form fringes, as shown in Figure 2.16. The center of the fringes define the location

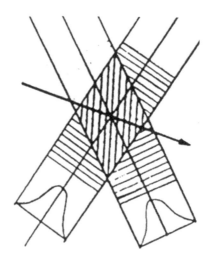

FIGURE 2.16
Formation of fringes by two identical laser beams.

of the point of measurement in a fluid. As the fluid particles pass through these fringes, there is a Doppler effect. The frequency of the Doppler shift that takes place is found to be proportional to the velocity of the fluid particles.

In a fringe LDV, the signals captured from the passing of photon particles through the fringes are autocorrelated. The correlator data is subjected to statistical analysis using the Fourier cosine transform that eventually produces a velocity distribution where a peak can be identified that signifies the velocity of the flow at the point of measurement perpendicular to the fringes. Bragg cells or a frequency shifting device can also be used to determine the direction of the flow. A typical representation of correlator data and the final velocity distribution are shown in Figure 2.17.

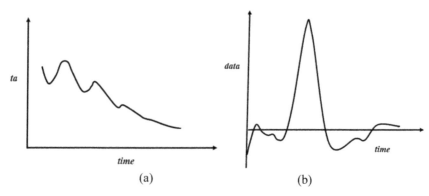

FIGURE 2.17
Typical form of (a) correlation data and (b) velocity distribution of fringe LDV system.

We can explain the underlying principle of a fringe LDV using the following two simple expressions [102]:

$$s = \frac{\lambda}{2\sin\left(\dfrac{\alpha}{2}\right)}$$

$$f_D = \left[\frac{2V\sin\left(\dfrac{\alpha}{2}\right)}{\lambda}\right]$$

where,
λ is the wavelength of the laser
α is the angle between the two converging beams
V is the velocity
f_D is the Doppler frequency

Say, for example, we want to measure velocity up to a maximum value of 100 m/s and we have a correlator whose maximum frequency capability is 20 MHz. Then the shortest fringe crossing time would be the inverse of 20 MHz or $100 \times 50 \times 10^{-9}$ seconds, giving the minimum fringe spacing to be $100 \times 50 \times 10^{-9}$ meters or 5 micrometers. By choosing a laser of a particular wavelength, λ, and by adjusting the angle, α, the required fringe spacing can be obtained.

Figure 2.18 shows the front section of a three-dimensional LDV designed by Ahmed et al. [103] to work in back scatter mode. In this design, the two equal laser lights are crossed on the mechanical axis of the probe head. By

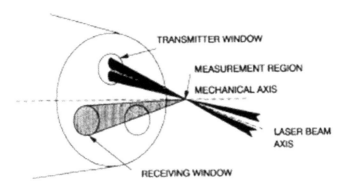

FIGURE 2.18
Front head view of the three-dimensional fiber optic laser probe head and the formation of the fringes. (After: Ahmed, Elder, Foster, and Jones, 1990 [103].)

rotating the probe head at three known angles, the velocity components in all three directions can be obtained.

Further details of the various components used in the probe head of this three-dimensional fringe LDA [103] are given in Figure 2.19. A block diagram of the overall LDA system developed is shown in Figure 2.20.

In Figure 2.19, laser light from the fiber is from a light source is fed into the probe head using a single-mode fiber. The laser light from the fiber is passed through a plano-convex lens to create the laser beam waist at a pre-determined distance. In fact two waists are required to form fringes. This is done with the help of a beam-splitter. The laser light exiting the lens is made obliquely incident on one end of the beam-splitter. The beam-splitter helps split the laser beam into two identical beams that come out at its other end at converging angles. The two beams thereafter recombine to form fringes at the waists. Using a mirror, the fringes are then re-positioned at an angle on the mechanical axis of the probe at the point marked as "M." The light scattered by particles as they pass through this point are collected obliquely through a second mirror and using a biconvex lens focussed on the end of a multi-mode fiber.

2.2.5.2 Time of Flight or Laser Two Focus System

The basic principle behind the time of flight or L2F system [104] involves the determination of the time it takes for the fluid particles to cross the waists of the two identical laser beams. The laser beam from a laser source is split into two identical beams, but unlike the LDV system, instead of crossing, the beams are placed parallel to one another at a fixed distance between the waists of the two beams. By knowing the distance and the time it takes for the fluid particles to cross the waists of the two beams, the velocity of the

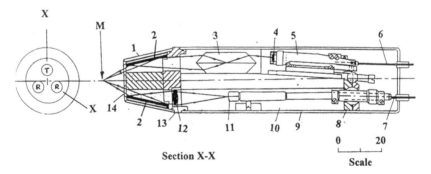

Section X-X

FIGURE 2.19
Internal details of various components of the LDV. (After: Ahmed, Elder, Foster, and Jones, 1990 [103].) Keys: 1-Front Cover; 2-Mirror; 3-Beam-splitter; 4-Transmitter lens; 5-Transmitter module; 6-Single-mode fiber; 7-Multi-mode fiber; 8-Bearing; 9-Cover; 10-Main body; 11-Pinhole; 12-Receiver lens; 13-Reciever lens module; 14-Window; T-Transmitter window; R-Receiver window.

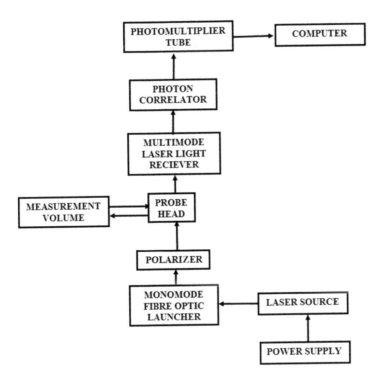

FIGURE 2.20
Block diagram of a fiber optic LDA system.

flow at the point of measurement can be determined. So the task boils down to the determination of the time. This is achieved by statistical means. In L2F, instead of auto-correlation, cross-correlation of the correlator data registered on the correlator is performed.

Figure 2.21 (a) shows particles passing through the waists or spots and Figure 2.21 (b) shows the determination of the time between the two peaks of the velocity distribution during their passage.

A histogram of the cross-correlator data helps to obtain the mean velocity and mean angle of the flow at the measurement location, as shown in Figures 2.21 (c) and (d).

The design of a time of flight incorporating fiber optics has also been reported by Ahmed et al. [105]. The process is very similar to that of the fiber optic laser, LDV, described earlier. In this design the light from the laser source was launched at the measurement volume using a monomode fiber. Multimode fiber was used to collect the scattered signals, which were amplified and processed electronically.

Figure 2.22 shows the details of this design.

Although laser anemometry is now an important tool in fluid mechanical investigation of complex flows [106], further useful guidance is provided in

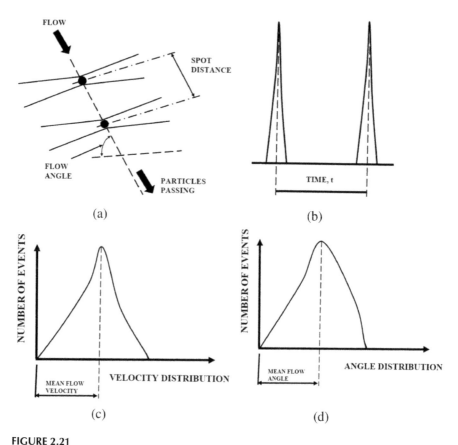

FIGURE 2.21
(a) Basic working principle of L2F and (b) time determination between particles, passing through the waists; (c) mean velocity determination from velocity distribution; (d) mean angle determination from angle distribution.

the book of Durst et al. [100], the review of Buchhave et al. [107], and the discussion of errors by Gould and Loseke [108].

2.2.6 Particle Image Velocimetry (PIV)

PIV is a later invention compared to LDV or L2F and has its origin in what is often referred to as Laser speckle velocimetry, a technique that was developed in the late 1970s. By the early 1980s it was realized that individual particles could be tracked if the particle concentration in a flow was reduced to certain levels and the flows could be split into very small areas of "interrogation" and obtain the velocity for each area [109, 110]. With the availability of advanced recording capabilities by digital camera and high computing power, the technique has been developing rapidly [111–115].

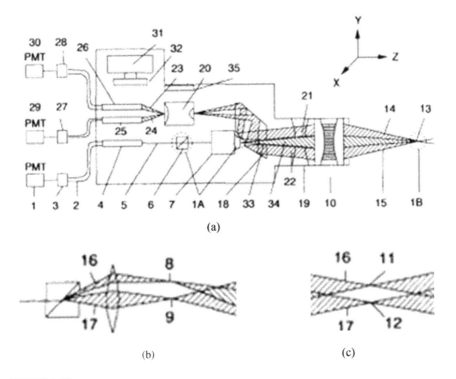

(a)

(b)　　　　　　　　　　　(c)

FIGURE 2.22
(a) General arrangement of a L2F design; (b) details of 1A; (c) details of 1B. (After: Ahmed, Elder, Foster, Jones, and Tatum, 1992 [105].) Keys: 1-Laser source; 2-Monomode fiber; 3-Launcher; 4-Microlens; 5-Laser beam; 6-Prism; 7-Microscope objective; 8,9-Waists imaged to form measurement volume; 10-Lens assembly; 11,12-Waists at measurement volume; 13-Measurement volume; 14,15-Received beams; 16,17-Transmitted beams; 18-Prism; 19-Lens; 20-Magnifier; 21,22,23-Received beams; 25,26-Monomode fiber; 27,28-Multimode fiber holder; 29,30-Photomultiplier tube; 31-Stepper motor; 32-Worm gear; 33-Transmitted beam; 34-Metal cone; 35-Bearings.

We can explain the principle of PIV in the following simple manner. Suppose two images of the same fluid particles are captured at two intervals of time. If we can track the distance that each particle has traversed during that time, then dividing the distance by time would give the velocity of each particle within an area of interest. This task is performed by using statistical means of conducting auto-correlation of each particle at the two time instances and cross-correlating with other particles at the same.

A typical PIV apparatus consists of a digital camera, a strobe, and a synchronizer. The camera records the image, the strobe illuminates the flow, and the synchronizer activates the whole system. A fiber optic cable or liquid light guide may also be used connect the laser to the lens setup. Figure 2.23 shows a simple PIV setup.

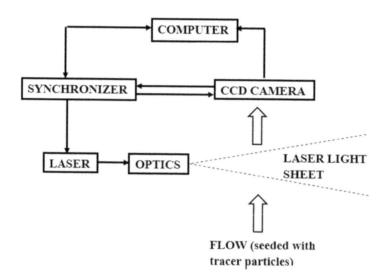

FIGURE 2.23
Basic PIV system setup.

Particle image velocimetry methods do not, in general, measure velocity components along the z-axis. Parallax errors may also be introduced in the velocity measurements in the x and y directions. To overcome these problems, stereoscopic PIV, which uses two cameras, may be used to measure all three velocity components. Other complex PIV setups that may be used include dual plane stereoscopic PIV, multi-plane stereoscopic PIV, holographic PIV, thermographic PIV, and so forth.

In PIV measurements, velocity vectors are obtained by cross-correlating the intensity distributions over small areas of the flow field to produce spatially averaged representations of the actual velocity field. This often impacts the accuracy of the subsequent spatial derivatives of the parameters such as the vorticity or spatial correlation function. The discussion of PIV errors by Lourenco and Krothapalli [116] are instructive.

2.3 Reduction of Data and Analysis

2.3.1 Theoretical Derivation of the Pressure Coefficient of a Jet Sheet

Figure 2.24 shows the geometry of a jet bending on a Coanda surface (bound surface) on one side and the other being free to atmosphere. Based on the works of Roderick [117] and Korbacher [118], the pressure coefficient of the jet for incompressible and compressible jets are provided. The notations and assumptions used are also summarized below.

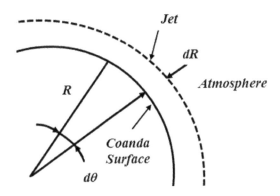

FIGURE 2.24
Geometry of jet on a Coanda surface.

NOTATIONS

p_S:	Coanda surface pressure
p_∞:	ambient pressure
p:	total pressure
Δp:	pressure difference
C_p:	pressure coefficient
$C_{Pincomp}$:	incompressible pressure coefficient for thin jet
C_{Pcomp}:	compressible pressure coefficient for thin jet
$C_{P'incomp}$:	incompressible pressure coefficient for thick jet
F_C:	centrifugal force
F_p:	pressure force
R:	radius curvature of jet
W:	jet width
b_{jet}:	jet thickness
U_{jet}:	jet velocity
TH:	thrust per jet unit width
ρ_∞:	ambient density
$d\theta$:	elemental angle
b_{jet}/W:	aspect ratio of the jet
U_∞:	free stream velocity
M_∞:	free stream Mach number

2.3.1.1 Assumptions

- The jet is thin.
- The aspect ratio of the jet is high.
- The jet thickness is very small compared to the radius of curvature of the Coanda surface.

- The jet momentum is constant along the jet sheet.
- The pressure gradient across the jet sheet is negligible.
- The flow is ideal (inviscid and irrotational).

2.3.1.2 Coefficient of Pressure for Incompressible Flow

The centrifugal force acting on a flow element is:

$$F_C = \rho_\infty R \; d\theta dR \frac{U_\infty^2}{R}$$

The pressure force acting on a flow element is:

$$F_P = R \; d\theta dP$$

If the centrifugal force and pressure force acting on a flow element are in equilibrium, then:

$$dp = \rho_\infty dR \frac{U_\infty^2}{R}$$

Thus, giving:

$$\rho_\infty R \; d\theta dR \frac{U_\infty}{R} = R d\theta dP$$

With βjet $\approx dR$,

$$dp = \rho_\infty \beta_{jet} \frac{U_\infty^2}{R}$$

Across the curved jet sheet, the pressure difference:

$$\Delta p = -\left(P_S - P_\infty\right)$$

With $\Delta P \approx dP$,

$$\Delta p = -\rho_\infty dR \frac{U_\infty^2}{R}$$

The thrust per unit of jet sheet width:

$$T_H = \rho_\infty \beta_{jet} U_\infty^2$$

We can also express ΔP as:

$$\Delta p = -\frac{T_H}{R}$$

The standard definition of coefficient of pressure:

$$C_P = \frac{\Delta P}{\frac{1}{2}\rho_\infty U_\infty^2}$$

or,

$$C_P = -\frac{2b_{jet}}{R}$$

Thus, for a given thin jet sheet thickness, the pressure coefficient is independent of the pressure ratio.

2.3.1.3 Coefficient of Pressure for Compressible Flow

For compressible flow, we have to find an appropriate expression for the density term which is no longer constant. Consequently,

$$\Delta p = -\frac{2b_{jet}}{R}\frac{1}{2}\rho_\infty U_\infty^2$$

$$\frac{1}{2}\rho_\infty U_\infty^2$$

For an ideal flow, we can use the isentropic relations, which gives us:

$$C_P = \frac{\Delta P}{\frac{1}{2}\rho_\infty U_\infty^2}$$

$$= \frac{2}{\gamma M_\infty^2}\left[\frac{P_T}{P_S} - 1\right]$$

$$= \frac{2}{\gamma M_\infty^2}\left[\left(1 + \frac{\gamma-1}{2}M_\infty^2\right)^{\frac{\gamma}{\gamma-1}} - 1\right]$$

2.3.2 Determination of Pressure Coefficient from Static Pressure Port Data

The pressure coefficient at any port point, n, is given by:

$$C_P = \frac{p_n - p_\infty}{\frac{1}{2}\rho_\infty U_\infty^2}$$

In physical experimentation, it is worth noting that we can only measure pressure (static or total) differentially, that is with reference to another pressure. In other words, in the laboratory, the readings you obtained were not the direct values of:

$$p_n, p_\infty, p_T$$

so that:

$$\Delta p_n = p_{\text{ref}} - p_n$$

$$\Delta p_\infty = p_{\text{ref}} - p_\infty$$

$$\Delta p_T = p_{\text{ref}} - p_T$$

and:

$$C_{Pn} = \frac{\Delta p_n - \Delta p_\infty}{\Delta p_T - \Delta p_\infty}$$

Generally, the reference pressure is the ambient or free stream pressure, so that:

$$p_{\text{ref}} = p_\infty$$

$$\Delta p_{\text{ref}} = \Delta p_\infty$$

2.3.3 Lift and Drag Coefficients from Surface Pressure Coefficient Distribution

It is customary to drill several tapping points on the top and bottom surfaces of an airfoil. For a subsonic airfoil, more points are allocated towards the leading edge than the trailing edge because the pressure coefficients are expected to be changing more rapidly at the leading edge with changes in the angle of incidence. Figure 2.25 is a schematic of an airfoil on which 20 pressure holes have been drilled. Pressure tubes from each hole can then be connected to a multi-tube manometer.

The steps, thereafter, to be followed are as follows:

- Calculate C_p at each static pressure tap for each incidence and enter in a single table.
- Plot C_p distribution against both (x/c) and (y/c).
- Graphically evaluate C_N and C_T at each incidence angle (suffix N and T denote normal and tangential directions to chord line, respectively). Use the following formulae:

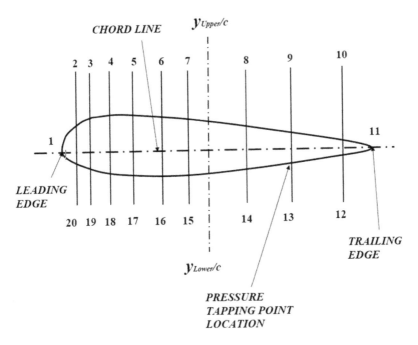

FIGURE 2.25
Tapping points on an airfoil.

$$C_N = -\int_0^1 \left(C_{P_{Upper}} - C_{P_{Lower}} \right) d\left(\frac{x}{c} \right)$$

$$C_T = \int_{\frac{y_{lower}}{c}}^{\frac{y_{Upper}}{c}} C_p d\left(\frac{x}{c} \right)$$

2.3.3.1 Lift Coefficient

The lift coefficient can be obtained using:

$$C_L = C_N \cos\alpha - C_T \sin\alpha$$

Generally, $C_N > C_T$, and the angle of incidence is small. Then the following approximation may used:

$$C_L \approx C_N \cos\alpha$$

2.3.3.2 Drag Coefficient

The drag coefficient can be expressed as:

$$C_D = C_N \sin\alpha + C_T \cos\alpha$$

Since, C_D, is expected to be small, further approximation to the above expression is not advisable. Furthermore, the drag coefficient obtained from the above method captures only the pressure drag, and not the total drag.

2.3.4 Determination of the Profile Drag by the Wake Traverse Method

Wake traverse method can be used to determine the profile drag on a body. The principle behind the method can be explained as follows. The presence of the body alters the flow field. The changes are more significant downstream of the airfoil than upstream in a subsonic flow. The changes are caused by a loss of momentum of the fluid downstream of the body.

To determine the loss in momentum, we need to determine the density and velocity changes between the upstream and downstream flows about the body. Assuming that the density changes are negligible for most subsonic flows, we only need to find changes in velocity or the momentum deficit in the wake of the body. This can be done from a consideration of the velocity distribution both upstream and downstream of the airfoil using a variety of methods.

We will describe here the most commonly used method, which employs static and total pressure measurements using pressure probes. The velocity distributions upstream of an airfoil, placed in a wind tunnel, are expected to be uniform, and often a single velocity measurement of a static-total head probe upstream of the airfoil may be sufficient to produce the velocity distribution.

In the wake, however, the velocity distribution will not be uniform and measurements at several points would be required to obtain the velocity profile distribution at a measurement plane of the wake. In this situation, we can use a rake, which is made up of a number of thin total head measuring probes and static pressure measuring probes. Because of losses in a wake, the velocity cannot be accurately determined because of uncertainties in both the total and static measurements.

The problem could be avoided if we could take measurements at a distance, far downstream, where the static pressure recovers to become the same as that of the upstream, so that the loss in the total pressure provides the magnitude of the velocity reduced in the wake. But because of viscosity, the wake begins to fade rapidly in the downstream direction making it difficult to detect the velocity reduction in the wake. A practical solution, therefore, is to take measurements at a location about a quarter chord distance downstream of the airfoil and apply some corrections to the results obtained.

2.3.4.1 Theoretical Approach

We will now describe an approach that blends theoretical analysis and practical approach to produce a useful method for fluid mechanical investigation. The basis of this method has its origin in the works of Jones [119] and details of the procedure are given by Ahmed [120].

Consider the control volume ACGE around the airfoil as shown in Figure 2.26. The fluid is assumed to be incompressible, i.e., density to be constant or negligible:

Mass flow crossing AC (top) and EG (bottom) $\int_A^E \rho_\infty (U_\infty - u) dz$
Consider **momentum**:

Momentum out over CG $= \int_C^G \rho_\infty u^2 dz$

Momentum in over AE $= \int_A^E \rho_\infty U_\infty^2 dz$

Momentum out AC and $GE = \int_C^G \rho_\infty (U_\infty - u) U_\infty dz$

Force acting on fluid$= -D$
Since the rate of momentum increase equals the sum of all external forces, we get:

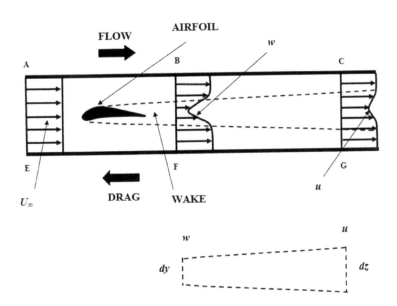

FIGURE 2.26
Determination of drag by wake traverse method.

$$-D = \int_C^G [\rho_\infty u^2 - \rho_\infty U_\infty^2 + \rho_\infty (U_\infty - u)U_\infty] dz$$

Simplifying, $D = \int_C^G \rho_\infty u (U_\infty - u) dz$
But

$$C_D = \frac{D}{\frac{1}{2}\rho_\infty U_\infty^2 c}$$

Hence, in the wake:

$$C_D = 2\int \frac{u}{U_\infty}\left(1 - \frac{u}{U_\infty}\right)d\left(\frac{z}{c}\right)$$

2.3.4.2 Practical Approach

In practice, $Y_1 Y_2$ is taken close to the airfoil, where local static pressure has not yet recovered to the free stream static pressure value. So, we measure total and static pressure at $Y_1 Y_2$ instead of at $Z_1 Z_2$. Outside of the wake, the total pressure is assumed constant; in other words, we have a situation similar to a potential flow. Now, in order to apply Bernoulli's equation, we make the following assumption: the total pressure does not vary within the wake. Remember also that we have assumed incompressible flow. Then in the wake,

$$H = p_\infty + \frac{1}{2}\rho_\infty u^2 = p + \frac{1}{2}\rho_\infty w^2$$

From *continuity*, with constant density,

$$w \cdot dz = \cdot u \cdot dy$$

Therefore,

$$D = \int_{y_1}^{y_2} \rho_\infty w (U_\infty - u) dz$$

But

$$u^2 = \frac{H - p_\infty}{\frac{1}{2}\rho_\infty}; \; w^2 = \frac{H - p}{\frac{1}{2}\rho_\infty}$$

Also, by definition

$$C_D = \frac{D}{\frac{1}{2}\rho_\infty U_\infty^2 c}$$

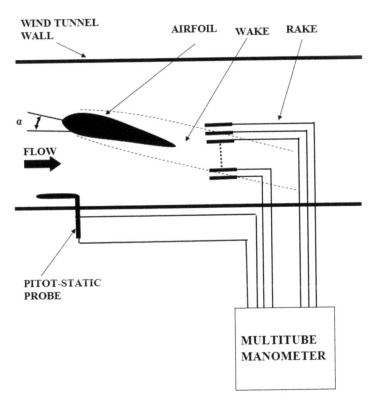

FIGURE 2.27
Test setup for wake traverse experiment.

and with

$$g = \frac{u^2}{U_\infty^2} = \frac{H - p_\infty}{\frac{1}{2}\rho_\infty U_\infty^2}$$

$$C_D = \int C_d d\left(\frac{y}{c}\right) = \int 2\sqrt{g - C_p}\left(1 - \sqrt{g}\right) d\left(\frac{y}{c}\right)$$

Figure 2.27 shows the test setup generally used in profile drag determination.

2.3.4.3 *Data Reduction and Plotting*

- Calculate g and C_p for each position in the rake, where

$$g = \frac{H - p_\infty}{\frac{1}{2}\rho_\infty U_\infty^2}$$

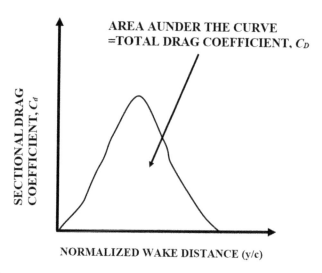

**AREA AUNDER THE CURVE
=TOTAL DRAG COEFFICIENT, C_D**

SECTIONAL DRAG COEFFICIENT, C_d

NORMALIZED WAKE DISTANCE (y/c)

FIGURE 2.28
Qualitative plot of C_d versus y_{wake}/c.

$$C_P = \frac{p - p_\infty}{\frac{1}{2}\rho_\infty U_\infty^2}$$

- Calculate and enter the product C_d (sectional drag coefficient) where

$$C_d = 2\sqrt{g - C_p}\left(1 - \sqrt{g}\right)$$

- Thereafter, graphically determine the total drag coefficient, C_D where

$$C_D = \int C_d d\left(\frac{y_{wake}}{c}\right)$$

A typical plot of C_d versus y_1/c (here y_1 is y_{wake}) is given in Figure 2.28.

2.4 Concluding Remarks

We are still a long way off from claiming that we can model complex flow phenomena with absolute confidence. Physical experimentation is, therefore, a necessary prerequisite on which to build confidence. Most flows of practical importance, such as Coanda flows, are complex, turbu-lent, and highly three-dimensional. Investigators, therefore, will have to consider carefully the pros and cons before employing any numerical or experimental method or some combination of them. Consequently, it is almost inevitable that Coanda flow will rely on some degree of empiricism, which in turn would require an excessive margin of safety while evaluating the unknowns.

References

1. Ahmed, N.A., (2014), Modern and conventional tools for effective teaching and learning experiences in aerospace engineering, Chapter 3, in F. Alam (ed.), *Using Technology Tools to Innovate Assessment, Reporting, and Teaching Practices in Engineering Education*, published by IGI Global (www.igi-global.com), 2014, pp. 21–39.
2. Roger Temam, R., (1984), *Navier–Stokes Equations: Theory and Numerical Analysis*, ACM Chelsea Publishing.
3. Batchelor, G.K., (1967), *An Introduction to Fluid Dynamics*, Cambridge University Press, Cambridge.
4. White, F.M., (2006), *Viscous Fluid Flow*, McGraw-Hill, Boston.
5. Smits, A.J., (2014), *A Physical Introduction to Fluid Mechanics*, John Wiley and Sons, Inc.
6. Schlichting, H., Gersten, K., Krause, E., Oertel, H., Jr., and Mayes, C., (2004), *Boundary-Layer Theory*, 8th edition, Springer.
7. Gad-el-Hak, M., (2000), *Flow Control-Passive, Active, and Reactive Flow Management*, 1st edition, Cambridge University Press.
8. Tennekes, H., and Lumley, J.L., (1972), *A First Course in Turbulence*, The MIT Press, Cambridge, MA.
9. Houghton, E.L., and Boswell, R.P., (1969), *Further Aerodynamics for Engineering Students*, 1st edition, Edward Arnold (Publishers) Ltd, London.
10. Kuptsov, L.P., (2001), Einstein rule, in *Hazewinkel, Michiel, Encyclopedia of Mathematics*, Kluwer Academic Publishers.
11. Panton, R.L., (1996), *Incompressible Flow*, 2nd edition, John Wiley and Sons, Inc.
12. Reynolds, O., (1895), On the dynamical theory of incompressible viscous fluids and the determination of the criterion, *Philosophical Transactions of the Royal Society of London. A*, vol. 186, pp. 123–164.
13. Bradshaw, P., (1999), The best turbulence models for engineers, in *Modeling Complex Turbulent Flows*, M.D. Salas, J.N. Hefner, L. Sakell (eds.), vol. 7, of the series ICASE/LaRC Interdisciplinary Series in Science and Engineering, Springer, Dordrecht.
14. Von Karman, T., (1921) (in German), 'Uber laminare and turbulente', Reiburg Zeitschrift fur angewandte Mathematic und Mechanic, Germany.
15. Pohlhausen, K. (1921) (in German), 'Zur naherungsweisen Integration der Differential-gleichung der laminaren Greenzschicht', Zeitschrift fur angewandte Mathematic und Mechanic, Germany.
16. Walz, A., and Joerg, O.H., (1970), *Boundary Layers of Flow and Temperature*, The MIT Press, Cambridge, MA.

17. Patel, V.C., (1974), A simple integral method for the calculation of thick axisymmetric turbulent boundary layers, *The Aeronautical Quarterly*, vol. 25, no. 1, pp. 47–58.
18. Hinze, J.O., (1975), *Turbulence*, 2nd edition, McGraw-Hill, New York.
19. Kays, W.M., and Crawford, M.E., (1993), *Convective Heat and Mass Transfer*, 3rd edition, McGraw-Hill, New York.
20. Ludweig, H., and Tillman, W., (1950), Experimental determination of skin friction coefficient in a turbulent boundary layer with a longitudinal pressure gradient, NACA TM 1285 (translation).
21. Head, M.R., (1958), Entrainment in the turbulent boundary layer, Aeronautical Research Council R & M 3152.
22. Bradshaw, P., (1969), Effects of streamline curvature on turbulent flow, AGARDograph no. 169, AGARD 7 Rue Ancelle, 92200 Neuille Sur Seine, France.
23. Howarth, L., (1951), The boundary layer in three-dimensional flow, part I: derivation of the equations of flow along a general curved surface, *The London, Edinburgh, and Dublin Philosophical Magazine and Journal of Science* , vol. 42, no. 326, pp. 239–243 (Series 7).
24. Patel, V.C., and Sotiropoulos, F., (1997), Longitudinal curvature effects in turbulent boundary layers, *Progress in Aerospace Science*, vol. 33, pp. 1–70.
25. Lamb, H., (1945), *Hydrodynamics*, 6th edition, Dover Press, New York.
26. Squire, H.B., (1965), The growth of a vortex in a turbulent flow, *The Aeronautical Quarterly*, vol. 16, no. 3, pp. 302–306.
27. Westwater, F.L., (1935), Rolling up of the surface of discontinuity behind an aerofoil of finite span, Aeronautical Research Council R & M no. 1962.
28. Moore, D.W., (1974), A numerical study of the roll up of a finite vortex sheet, *Journal of Fluid Mechanics*, vol. 63, pp. 225–235.
29. Tani, I, (1977), History of boundary layers, *Annual Review of Fluid Mechanics*, vol. 9, pp. 87–111.
30. Rae, W.H., Jr., and Pope A., (1984), *Low-Speed Wind Tunnel Testing*, 2nd edition, John Wiley & Sons, New York.
31. Pope, A., and Goin, K.L., (1999), *High-Speed Wind Tunnel Testing*, 3rd edition, John Wiley & Sons, New York.
32. Chen, Z., He Yi, S., Ahmed, N.A., Ping Kong, X., and Chen Quan, P., (2015), Transient behaviour of shock train with or without controlling, *Experimental Thermal and Fluid Science*, vol. 66, pp. 79–96.
33. Flynn, T.G., and Ahmed, N.A., (2005), Investigation of rotating ventilator using smoke flow visualisation and hot-wire anemometer, in *Proceedings of 5th Pacific Symposium on Flow Visualisation and Image Processing*, Whitsundays, Australia, Paper No. PSFVIP-5-214.
34. Findanis, N., and Ahmed, N.A., (2008), The interaction of an asymmetrical localised synthetic jet on a side supported sphere, *Journal of Fluids and Structures*, vol. 24, no. 7, pp. 1006–1020.
35. Behfarshad, G., and Ahmed, N.A., (2011), Vortex flow asymmetry of slender delta wings, *International Review of Aerospace Engineering*, vol. 4, no. 3, pp. 184–188.
36. Behfarshad, G., and Ahmed, N.A., (2011), Experimental investigations of sideslip effect on four slender delta wings, *International Review of Aerospace Engineering*, vol. 4, no. 4, pp. 189–197.

37. Pitot, H., (1732), Description d'une machine pour mesurer la vitesse des eaux courantes et le sillage des vaisseaux, Histoire de l'Académie royale des sciences avec les mémoires de mathématique et de physique tirés des registres de cette Académie pp. 363–376.

38. Darcy, H., (1858), Note relative à quelques modifications à introduire dans le tube de Pitot, Annales des Ponts et Chaussées, pp. 351–359.

39. By Chaos386 – Own work, CC BY-SA 3.0, https://commons.wikimedia.org/w/index.php?curid=3301685

40. Public Domain, https://commons.wikimedia.org/w/index.php?curid=3938571

41. Anderson, J.D., Jr., (2005), *Fundamentals of Aerodynamics*, 4th edition, McGraw-Hill, Series in Aeronautical and Aerospace Engineering.

42. Rashid, D.H, Ahmed, N.A., and R.D.Archer, (2003), Study of aerodynamic forces on rotating wind driven ventilator, *Wind Engineering*, vol. 27, no. 1, 2003, pp. 63–72.

43. Pisasale, P., and Ahmed, N.A., (2002), Theoretical calibration of a five hole probe for highly three dimensional flow, *International Journal of Measurement Science and Technology*, vol. 13, pp. 1100–1107.

44. Pisasale, P., and Ahmed, N.A., (2002), A novel method of extending the calibration range of five hole probe for highly three dimensional flows, *Journal of Flow Measurement and Instrumentation*, vol. 13, nos. 1–2, pp. 23–30.

45. Pissasale, P., and Ahmed, N.A., (2003), Examining the effect of flow reversal on seven-hole probe measurements, *AIAA Journal*, vol. 41, no. 12, pp. 2460–2467.

46. Pissasale, P., and Ahmed, N.A., (2004), Development of a functional relationship between port pressures and flow properties for the calibration and application of multi-hole probes to highly three-dimensional flows', *Experiments in Fluids*, vol. 36, no. 3, pp. 422–436.

47. Vagt, J.D., and Fernholz, H.H., (1979), A discussion of probe effects and improved measuring techniques in the near wall region of an incompressible three-dimensional turbulent boundary layer, AGARD CP 271.

48. Fernholz, H.H., and Vagt, J.D., (1981), Turbulence measurements in an adverse-pressure-gradient three-dimensional turbulent boundary layer along a circular cylinder, *Journal of Fluid Mechanics*, vol. 111, pp. 233–269.

49. Bertelrud, A., (1977), Total head/static measurements of skin friction and surface pressure, *AIAA Journal*, vol. 15, pp. 436–438.

50. Rechenberg, I., (1963), Messung der turbulenten wandschubspannung, Z *Flugwiss*, vol. 11, pp. 429–438.

51. Winter, K.G., (1977), An outline of the techniques available for the measurement of skin friction in turbulent boundary layers, *Progress in Aerospace Science*, vol. 18, pp. 1–57.

52. Haritonidis, J.H., (1989), The measurement of wall shear stress, in M. Gad-el-Hah (ed.), *Advances in Fluid Mechanics and Measurements*, Lecture notes in Engineering, Chapter 6, vol. 45, Springer, Berlin, Hedelberg, pp. 229–261.

53. Hanratty, T.J., and Campbell, J.A., (1996), Measurement of wall shear stress, in R. Goldstein (ed.), *Fluid Mechanics Measurements*, 2nd edition, Taylor & Francis Group, Routledge, pp. 575–640.

54. Preston, J.H., (1953), The determination of turbulent skin friction by means of pitot tubes, *Journal of Royal Aeronautical Society*, vol. 58, pp. 109–121.

55. Stanton, T.E., Marshall, D., and Bryant, C.W., (1920), On the condition at the boundary of a fluid in turbulent motion, *Proceedings of the Royal Society of London, Series A*, vol. 97, pp. 413–434.

56. Konstantinov, N.I., (1953), Comparative investigation of the friction stress on the surface of a body, *Energomashinostroenie*, vol. 176, pp. 201–213 (Translated 1960 DSIR RTS 1499).

57. Chue, S.H., (1975), Pressure probes for fluid measurement, *Progress in Aerospace Science*, vol. 16, no. 2, pp. 147–223.

58. Nece, R.E., and Smith J.D., (1970), Boundary layer stress in rivers and estuaries, *Journal of Water Harbor Research*, vol. 96, pp. 335–358.

59. Ackerman, J.D., Wong, L., Ethier, C.R., Allen, D.G., and Spelt, J.K., (1994), Preston-static tubes for the measurement of wall shear stress, *Journal of Fluids Engineering*, vol. 116, pp. 645–649.

60. Lien, J., and Ahmed, N.A., (2006), Skin friction determination in turbulent boundary layers using multi-hole pressure probes, in 25th AIAA Applied Aerodynamics Conference, San Francisco, USA, AIAA-2006-3659, pp. 8–10.

61. Lien, J., and Ahmed, N.A., (2011), An examination of the suitability of multi-hole pressure probe technique for skin friction measurement in turbulent flow, *Journal of Flow Measurement and Instrumentation*, vol. 22, pp. 153–164.

62. Gatto, A., Byrne, K.P., Ahmed, N.A., and Archer, R.D., (2001), Pressure measure-ments over a cylinder in crossflow using plastic tubing, *Experiments in Fluids*, vol. 30, no. 1, pp. 43–46.

63. Gatto, A., Ahmed, N.A., and Archer, R.D., (2000), Investigation of the upstream end effect of the flow characteristics of a yawed circular cylinder,*The RAeS Aeronautical Journal*, vol. 104, no.1033, pp. 125–128, 253–256.

64. Gatto, A., Ahmed, N.A., and Archer, R.D., (2000), Surface roughness and free stream turbulence effects on the surface pressure over a yawed circular cylin-der, *AIAA Journal of Aircraft*, vol. 38, no. 9, pp. 1765–1767.

65. Bradshaw, P., (1971), *An Introduction to Turbulence and Its Measurement*, Pergamon Press, London.

66. Reynolds, A.J., (1974), *Turbulent Flows in Engineering*, John Wiley & Sons, New York.

67. Sarma, G.R., (1993), Analysis of a constant voltage anemometer circuit, *Instrument Measurement Technology Conference Proceeding*, vol. 1, pp. 731–736.

68. Ferreira, R.P.C., and Freire, R.C.S., (2001), Hot-wire anemometer with tem-perature compensation using only one sensor, *IEEE Transaction Instrument Measurement*, vol. 50, pp. 954–958.

69. Moreira,M A., Oliveira, A., Dorea, C.E.T., Barros, P.R., and Neto, J.S.R., (2008), Sensors characterization and control of measurement systems with thermo-resistive sensors using feedback linearization, *Instrument Measurement Technology Conference Proceeding*, vol. 1, pp. 2003–2008.

70. Martins, V.S.G., Freire, R.C.S., and Catunda, S.Y.C., (2012), Sensitivity analysis and automatic adjustment of a controlled-temperature thermo-resistive-based anemometer, *Instrument Measurement Technology Conference Proceeding*, vol. 1, pp. 1876–1880.

71. Stornelli, V., Ferri, G., Leoni, A., and Pantoli, L., (2017), The assessment of wind conditions by means of hot wire sensors and a modifed Wheatstone bridge architecture, *Sensors and Actuators A*, vol. 262, pp. 130–139.

72. Public domain: https://commons.wikimedia.org/wiki/File:Wheatstonebridg e.svg
73. Aydin, M., and Leutheusser, M.J.H., (1980), Very low velocity calibration and application of hot wire probes, *DISA Information*, publication number 25, pp. 17–18.
74. Gibbings, J.C., Madadnia, J., and Yousif, A.H., (1995), The wall correction of the hot wire anemometer, *Flow Measurement and Instrumentation*, vol. 6, no. 2, pp. 127–136.
75. Prat, R.L., and Bowsher, J.W., (1978), A simple technique in the calibration of hot wire anemometers at low air velocities, *DISA Information*, publication number 23, pp. 33–34.
76. Abdel-Rahman, A.A., (1995), On the yaw-angle characteristics of hot-wire anemometers, *Flow Measurement and Instrumentation*, vol. 6, no. 4. pp. 271–278.
77. Lee, T., and Budwig, R., (1991), Two improved methods for low-speed hot-wire calibration, *Measurement Science and Technology*, vol. 2, pp. 643–646.
78. Yue, Z., and Malmstrom, T.G., (1998), A simple method for low-speed hot-wire anemometer calibration, *Measurement Science and Technology*, vol. 9, pp. 1506–1510.
79. Johnstone, A., Uddin, M., and Pollard, A., (2005), Calibration of hot-wire probe using non-uniform mean velocity profiles, *Experiments in Fluids*, vol. 39, pp. 525–532.
80. Ozahi, E., Çarpinlioglu, M.O., and Gundigdu, M.Y., (2010), Simple methods for low speed calibration of hot-wire anemometers, *Flow Measurement and Instrumentation*, vol. 21, pp. 166–170.
81. Kohan, S., and Schwarz, W., (1973), Low speed calibration formula for vortex shedding from cylinders, *Physics of Fluids*, vol. 16, pp. 1528–1529.
82. Bruun, H.H., Farrar, B., and Watson, I., (1989), A swinging arm calibration method for low velocity hot-wire probe calibration, *Experiments in Fluids*, vol. 7, pp. 400–404.
83. Guellouz, M.S., and Tavoularis, S., (1995), A simple pendulum technique for the calibration of hot-wire anemometers over low-velocity ranges, *Experiments in Fluids*, vol. 18, pp. 199–203.
84. King, L.V., (1914), On the convection of Heat from small cylinders in a stream of fluid, *Philosophical Transaction of the Royal Society of London*, A 214, pp. 373–432.
85. Tsanis, I.K., (1987), Calibration of hot wire anemometers at very low velocities, *DANTEC Information*, publication number 23, pp. 13–14.
86. Johnson, F., and Eckelmann, H., (1984), A variable angle method of calibration for X probes applied to wall-bounded turbulent shear flow, *Experiments in Fluids*, vol. 2, pp. 121–130.
87. Lueptow R., Breuer, K., and Haritonidis, J., (1986), Computer aided calibration of X-probes using a look-up table, *Experiments in Fluids*, vol. , pp. 115–118.
88. Browne, L., Antonia, R., and Chua, L., (1988), Calibration of X-probes for turbulent flow measurements, *Experiments in Fluids*, vol. 7, pp. 201–208.
89. Willmarth, W.W., and Bogar, T.J., (1977), Survey and new measurements of turbulent structure near the wall, *Physics of Fluids*, vol. 20, pp. 9–21.
90. Maciel,Y., and Gleyzes, C., (2000), Survey of multi-wire probe data processing techniques and efficient processing of four-wire probe velocity measurements in turbulent flows, *Experiments in Fluids*, vol. 29, pp. 66–78.
91. Lecic, M.R., (2009), A new experimental approach to the calibration of hot-wire probes, *Flow Measurement and Instrumentation*, vol. 20, pp. 136–140.

92. Perry, A.E., and Morrison, G.L., (1971), Static and dynamic calibrations of constant-temperature hot-wire systems, *Journal of Fluid Mechanics*, vol. 47, pp. 765–777.
93. Li, W.Z., Khoo, B.C., and Xu, D., (2007), The thermal characteristics of a hot wire in a fluctuating freestream flow, *International Journal of Heat and Fluid Flow*, vol. 28, pp. 882–893.
94. Chew, Y.T., Shi, S.X., and Khoo, B.C., (1995), On the numerical near-wall corrections of single hot-wire measurements, *International Journal of Heat and Fluid Flow*, vol. 16, pp. 471–476.
95. Kovasznay, L.S.G., Kibens, V., and Blackwelder, R.F., (1970), Large-scale motion in the intermittent region of a turbulent boundary layer, *Journal of Fluid Mechanics*, vol. 41, p. 283.
96. White, A.D., and Rigden, J.D., (1962), Continuous gas maser operation in the visible, *Proceedings of IRE*, vol. 50, 1697.
97. Yeh, Y., and Cummins, H.Z., (1964), Localized fluid flow measurements with an He-Ne laser spectrometer, *Applied Physics Letters*, vol. 4, no. 10, pp. 176–178.
98. Foreman, J.W., George, E.W., and Lewis, R.D., (1965), Measurement of localized flow velocities in gases with a laser doppler flowmeter, *Applied Physics Letters*, vol. 7, no. 4, pp. 77–78.
99. Drain, L.E., (1980), *The Laser Doppler Technique*, John Wiley & Sons.
100. Durst, F., Melling, A., and Whitelaw, J.H., (1976), *Principles and Practice of Laser Doppler Anemometry*, Academic Press, London.
101. Melling, A., (1997), Tracer particles and seeding for particle image velocimetry, *Measurement Science and Technology*, vol. 8, no. 12, pp. 1406–1416.
102. Ahmed, N.A., (1999), Towards miniaturisation in laser velocimetry, key note address, in J. Sorio, and D. Honnery (eds), 2nd Australian Conference in Laser Diagnostic in Fluid Mechanics and Combustion Proceedings, Melbourne, pp. 28–35.
103. Ahmed, N.A., Elder, R.L., Foster, C.P., and Jones, J.D.C., (1990), Miniature laser anemometer for 3D measurements, *Journal of Measurement Science Technology*, vol. 1, pp. 272–276.
104. Ahmed, N.A., Elder, R.L., Foster, C.P., and Jones, J.D.C., (1991), Laser anemometry in turbomachines', *Proceedings of the Institution of Mechanical Engineers, Part G*, vol. 205, 1991, pp. 1–12.
105. Ahmed, N.A., Hamid, S., Elder, R.L., Foster, C.P., and Jones, J.D.C. and Tatum, R., (1992), Fiber optic laser anemometry for turbo machinery applications, *Optics and Lasers in Engineering*, vol. 15, nos. 2 and 3, 1992, pp. 193–205.
106. Ahmed, N.A., and Elder, R.L., (2000), Flow behaviour in a high speed centrifugal impeller passage under design and off-design operating conditions, *Fluids and Thermal Engineering, JSME International Series B*, vol. 43, no. 1, pp. 22–28.
107. Buchhave, P., George, W.K., and Lumley, J.L., (1979), The measurement of turbulence with the laser-Doppler anemometer, *Annual Review of Fluid Mechanics*, vol. 11, pp. 443–503.
108. Gould, R.D., and Loseke, K.W., (1993), A comparison of four velocity bias correction techniques in laser-Doppler velocimetry, *Journal of Fluids Engineering*, vol. 115, pp. 508–514.
109. Raffel, M., Willert, C., Wereley, S., and Kompenhans, J., (2007), *Particle Image Velocimetry: A Practical Guide*, Springer-Verlag.
110. Adrian, R.J., and Westerweel, J., (2011), *Particle Image Velocimetry*, Cambridge University Press.

111. Adrian, R.J., (1991), Particle-imaging techniques for experimental fluid mechanics, *Annual Review of Fluid Mechanics*, vol. 23, no. 1, pp. 261–304.
112. Adrian, R.J., (2005), Twenty years of particle image velocimetry, *Experiments in Fluids*, vol. 39, no. 2, pp. 159–169.
113. Santiago, J.G., Wereley, S.T., Meinhart, C.D., Beebe, D.J., and Adrian, R.J., (1998), A micro particle image velocimetry system, *Experiments in Fluids*, vol. 25, no. 4, pp. 316–319.
114. Fouras, A., Dusting, J., Lewis, R., and Hourigan, K., (2007), Three-dimensional synchrotron X-ray particle image velocimetry, *Journal of Applied Physics*, vol. 102, p. 064976.
115. Wereley, S.T., and Meinhart, C.D., (2010), Recent advances in micro-particle image velocimetry, *Annual Review of Fluid Mechanics*, vol. 42, no. 1, pp. 557–576.
116. Lourenco, L., and Krothapalli, A., (1995), On the accuracy of velocity and vorticity measurements with PIV, *Experiments in Fluids*, vol. 18, pp. 421–428.
117. Roderick, W.E.B., (1961), Use of the Coanda effect for the deflection of jet sheets over smoothely curved surface, Part II, University of Toronto, Institute of Aerophysics, Technical Note number 51.
118. Korbacher, G.K., (1962), The Coanda effect at deflection surfaces detached from jet nozzle, *Canadian Aeronautics and Space Journal*, vol. 8, no. 1, pp. 1–6.
119. Jones, B.M., (1937), The measurement of profile drag by the pitot-traverse method, The Cambridge University Aeronautics Laboratory (R & M No. 1688), Technical Report of the Aeronautical Research Committee for the Year 1935–1936, vol. I.
120. Ahmed, N.A., (2002), Implementation of a momentum integral technique for total drag measurement, *International Journal of Mechanical Engineering and Education*, vol. 30, no. 4, pp. 315–324.

3

Coanda Effect in Aeronautical Applications

3.1 Early Development of V/STOL Aircraft Using the Coanda Effect

The concept of V/STOL aircraft was implicit in the patent of Henri Coanda (Patent no. 1,108,652) as shown in Figure 1.1 of Chapter 1. The objective of the idea was to create acceleration of an airflow over a concave disc. During occupation of Paris in the Second World War, the German military became aware of Coanda's work and sought his help to design a large flying vehicle that could be powered by its latest jet engine technology. Accordingly, Coanda came up with the design of a disc-shaped vehicle that was 20 meters in diameter and required 12 004B large jet engines to power it. The year was 1944. At that point of the war, the Germans were losing and the Reich was fast collapsing. So Coanda's vehicle [1], as sketched in Figure 3.1, did not progress beyond the wind tunnel testing phase.

Post-war, Coanda's concept was revived by the Allied scientists, and in particular by Jack Frost in Canada. In Figure 3.2, Frost can be seen demonstrating [2] the Coanda effect by allowing pressurized air from the end of a tube to flow over the top of a metal disk. The flow curved down along the edges without separating and supported the disk in the air.

Frost was able to convince the US military to provide funding. By 1958, the AVRO Car VZ-9, which resembled a flying saucer [3], as shown in Figure 3.3, was built in Canada by Avro Aircraft, Ltd. as part of a secret, US military project.

From the onset, the development of Avrocar was marred by technical problems and desired performance was hardly achieved. The project was eventually cancelled in 1961. Although the Avrocar never became a reality as an operational vehicle, it nevertheless inspired many innovations, such as the technology of ducted fans. Today, a full-scale replica of the Avrocar can be seen as an exhibit at the Western Canada Aviation Museum in Winnipeg [4]).

The dream to integrate lift generation with propulsion remains as strong as ever, and interest in the Coanda effect is not diminishing. Most of the research efforts involve a combination of boundary layer control and circulation control. Various aeronautical conferences, forums, and workshops, such as those organized in 1986 and 2004 [5] on circulation control by the National

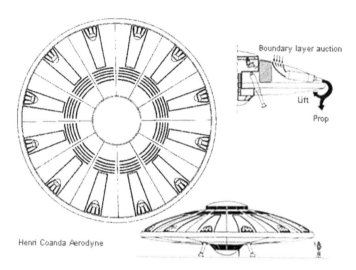

FIGURE 3.1
Coanda's design of an Aerodyne in 1944. (After: Coanda, 1935 [1].)

FIGURE 3.2
Picture of Jack Frost demonstrating the Coanda effect. (After: Arnodt [2].)

Aeronautics and Space Administration of the USA, are further testament to those efforts.

The rapid progress in the development of suitable power plants, such as the gas turbine with low specific weights (engine weight/static thrust), has given hope and further impetus to the development of aircraft requiring short runway lengths, and work continues toward the goal, continuing at full steam [6].

FIGURE 3.3
The first Avrocar at the Avro factory in 1958. (After: Bzuk, 2012 [3].)

This has resulted in the development of radical concepts of short take-off and landing (STOL), vertical take-off and landing (VTOL), or V/STOL aircraft as possible solutions. Unfortunately, aircraft produced thus far have poor performance during take-off and landing and high drag in cruise.

Various types of schemes have been proposed to achieve V/STOL performance. These include, amongst others:

- Compound aircraft, i.e., a helicopter with fixed wing
- Tilt-wing aircraft, which rotate the entire wing and propellers through 90 degrees while keeping the fuselage fixed
- Tilt-jets, tilt-propellers, or tilt-rotor aircraft, which rotate the thrust producers while maintaining the fuselage horizontal
- Fan-in-wing aircraft, which use fans submerged in the wings
- Jet powered aircraft, which use vectored thrust engines

The aircrafts seeking to exploit the Coanda effect, however, fall outside the types mentioned above and generally involve suction, blowing, or circulation control, or a combination of all three. It is, however, important to be aware of the aerodynamic concepts that define their distinguishing features.

3.2 Some Basic Aerodynamic Considerations

It is clear from the above discussions that high lift generation with low drag is an important driver of aerospace vehicle performance, and the Coanda effect can play a significant role in that narrative. To make sense of the process, the reader needs to have a greater appreciation for the flow mechanisms involved. We will, therefore, explore how lift is produced, how the production is associated with "circulation," and why "circulation control" studies are relevant.

These issues are not limited to aeronautic studies only. Suitable means of control and integration of lift production with a propulsive mechanism for marine vehicles are also important in hydronautics, and in many non-aeronautical applications, such as for performance enhancement of road vehicles, and so forth. Thus, the topics covered below will also be of benefit to any reader who may be interested in manipulating forces generated on moving bodies submerged in fluids.

3.2.1 General Description of Lift

When a body moves through air, it creates aerodynamic force around it. For convenience of analysis, the aerodynamic force can be resolved into drag and the lift force components. The drag force acts in the direction against the motion of the body, while the lift force acts in the direction perpendicular to the motion of the body. A schematic of forces acting on a moving airfoil is shown in Figure 3.4.

While lift force is pressure driven, drag force is viscous driven. It is customary to express these forces in their non-dimensional forms, that is, in terms of their lift and drag coefficients; and their relative importance is measured by the aerodynamic efficiency of the body. This aerodynamic efficiency, η_{aero}, is given by:

$$\eta_{aero} = Lift/\mathrm{Drag} = C_L/C_D$$

On a lifting body, such as on a wing, a low-pressure region is created on the top surface and a high-pressure region on the bottom surface. Since a force is

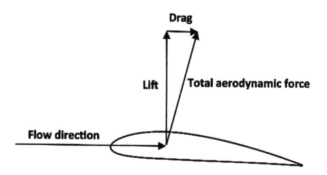

FIGURE 3.4
Aerodynamic force on an airfoil.

the product of pressure and area, the magnitude of the total lift force generated on a body is given by:

$$L = A\left(p_u - p_L\right)$$

where,
   L = Lift force (N)
   A = area (m²)
   p_u = pressure on the upper surface (N/m²) and
   p_L = pressure on the lower surface (N/m²)

From the above expression for L, the higher the suction pressure on the top
surface in relation to the pressure on the bottom surface, the higher is the lift
force acting on the body. In other words, to produce high lift, the pressure-
driven forces must dominate over viscous-driven forces and the flow must
remain attached to the body. This reinforces again why the Coanda effect
with its ability to prolong flow attachment has attracted so much attention
in aeronautics.

But not all body shapes can generate lift force of sufficient magnitude
that would be useful to an aeronautical engineer. We have already seen, for
example, in Figure 1.33 (a) of Chapter 1, that a square-shaped body would not
be suitable for lift production because this shape can very easily induce flow
separation at its corner edges and lead to high drag forces.

For a given aerodynamic force, the lift component can be increased if the
drag component is decreased. On this basis, Jones and his team at Cambridge
University in the 1930s proposed the concept of "streamlining" of a body.
This was a significant step that provided an effective means to increase a
body's lift-producing capability while reducing drag at the same time.

A streamlined body generally takes the shape of an airfoil. An airfoil is a
cross section of a wing and can be thought of as a wing of infinite span of
constant cross section. The infinite span aspect consideration implies that
the flow over an airfoil is two-dimensional because the three-dimensional
effects that originate from the flow breaking away from the wing tips and
the subsequent induced drag are absent. The lift coefficient of an airfoil
becomes greater than the lift coefficient of the whole wing and varies linearly with increases in the angle of incidence. This makes the mathematical
and experimental treatment of lift force simpler. However, for a particular
airfoil, beyond a certain angle of incidence, the viscous forces start to dominate over the pressure forces. This leads to rapid boundary layer growth and
eventually to flow separation. The airfoil then stalls and loses its ability to
produce lift further.

To prevent an airfoil or a wing stalling, some control measures, such as
boundary layer control, are introduced to ensure that the flow remains
attached to the wing surface longer.

3.2.2 Lift Generation Using Abstract Mathematical Concepts

Let us start with the best-case scenario and consider an "ideal" flow in which the aerodynamic efficiency of a lifting body is so high that in the limit it becomes infinity. This is possible when the drag force is zero or when the fluid motion suffers no losses. A corollary to this is that the total energy or the sum total of the potential and kinetic energies at every point in an ideal flow remains the same or becomes a constant.

The adoption of ideal flow assumption leads to some interesting mathematical concepts that have their origin in the 16th century. Although abstract in nature, these concepts are quite simple to understand and develop, and enable us to interpret effectively some important features of the associated flow mechanism.

3.2.2.1 The Concepts of Stream Functions and Stream Lines

To highlight some of the important flow features, it is sufficient to consider ideal two-dimensional flows. We will start with the abstract mathematical function of "stream function" to describe such flows since the "streamlines" that accompany this function are easier to comprehend as visual representation of paths traversed by fluid particles. Later we will see that we need another abstract function called the "potential function" to generate an airfoil shape.

The fact that a streamline is generated from a stream function makes it an abstract mathematical entity. There are several properties imposed on a streamline. One is that the velocity normal to a streamline at any point must be zero (Figure 3.5), implying that no flow can cross any streamline. This in turn makes the streamline a mathematical representation of a solid boundary where no flow crosses the boundary.

Stream functions are formulated to be expressed in units of m²/sec so that the velocity components at any point can be obtained by differentiating

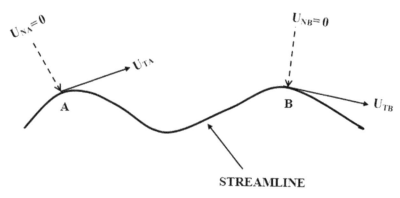

FIGURE 3.5
The concept of a streamline.

the stream functions with respect to their spatial coordinates directly. Once we have determined the velocity at a point, we can then proceed to apply the Bernoulli's principle and obtain the pressure at that point within the flow field.

A virtue of the ideal functions (stream and potential) is that we can simply add or subtract different functions to produce a new function and describe a new flow. For example, when we add the stream function of a uniform horizontal flow to the stream function of a uniform vertical flow, we create a new function that represents an inclined uniform flow. More on these mathematical expressions and how to derive them can be found in various fluid and aerodynamic books that are listed in the supplemental reading list.

3.2.2.2 Creating Body Shapes Using Stream Functions

A number of stream functions are singular in nature. This means that the flow generated by a singular function contains a point where the velocity is infinite or cannot be quantified. Such singular stream functions carry special significance in aeronautics because they can be combined to generate useful body shapes of determinable velocity and pressure fields.

In aeronautics, a useful stream function is that of a doublet, $\psi_{doublet}$. This function can be created by combining two singular functions of source and sink of equal strength and placing them so that the distance between them tends toward zero but does not actually become zero.

Let us see what happens when we place a doublet, $\psi_{doublet}$, in a free uniform stream, ψ_∞, where the stream is flowing from left to right. This will create a new function, say, ψ_C, that would be given by:

$$\psi_C = \psi_\infty + \psi_{doublet}$$

or

$$\psi_c = U_\infty . y - \frac{\mu}{2\pi r} \sin\theta$$

where

$$\psi_\infty = -U_\infty . y$$

$$\psi_{doublet} = \frac{\mu}{2\pi r} \sin\theta$$

U_∞ is the free stream velocity, and
μ is the strength of the doublet.

The flow field of the function, ψ_C, is shown in Figure 3.6.

By examining the flow field shown in Figure 3.6, we can observe a symmetric pattern of streamlines existing on both side of the doublet. On one side, the stream functions become more positive while on the other side, they

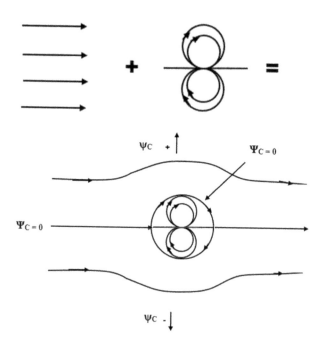

FIGURE 3.6
Flow over a doublet.

become more negative. This means the stream function at the center must have a value of zero. This streamline corresponding to this central stream function bifurcates and encircles the doublet before meeting again and continuing downstream.

The bifurcating streamlines encircling the doublet create the shape of a circle that has two stagnation points: one at the front and the other at the rear. The flow located inside the circle cannot escape outside of the it because the circle is a streamline. We can, therefore, ignore the inside flow altogether. The circle can now be considered to represent the cross section of a circular cylinder. In other words, we have created a streamline that is the mathematical equivalent of a two-dimensional stationary cylinder placed in a free stream, which is shown in Figure 3.7.

3.2.3 The Concepts of "Circulation" and "Vorticity"

The lift equation of Section 3.2.1 merely gives an expression to quantify the magnitude of the lift force, but it does not explain the mechanism of lift production. The existence of pressure differences between the top and bottom surfaces is not sufficient to produce a lift force on a body unless

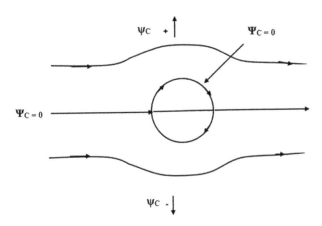

FIGURE 3.7
Flow over a cylinder.

the pressure difference information is communicated between the two surfaces. The mechanism by which this communication takes place is called "circulation."

Mathematically, circulation can be obtained from a line integral of a closed curve of a velocity field as given by the following expression:

$$\Gamma = \oint V \, dl$$

where Γ is circulation, V is the fluid velocity vector, and dl is the differential length vector of a small element of a closed curve.

Circulation, however, is related to rotation. This may appear confusing since rotation produces "vortex flow," or vorticity. And when vorticity is present, the flow no longer remains potential and is dominated by viscous forces. However, to generate lift, we need pressure forces to dominate. In other words, we need dominant pressure forces as well as rotation simultaneously to produce lift.

Let us explore how we can meet these conflicting requirements. The "vortex" motion creates vorticity with circulatory streamlines. Mathematically, this produces a flow field that can be potential (ideal) everywhere except at its center where the velocity becomes infinite. Because of the singularity, this flow field cannot exist physically. To find relevance to reality, we can imagine a nucleus to be present at its center, which would be rotating and thereby able to contain all the effects of vorticity. This nucleus would negate the singularity without altering the nature of the flow field around it. The circulatory streamlines of a vortex flow with a nucleus are shown in Figure 3.8.

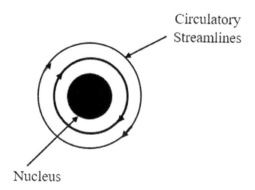

Circulatory
Streamlines

Nucleus

FIGURE 3.8
Circulation and vorticity.

3.2.4 Circulation on a Rotating Body (Rotating Circular Cylinder)

We can now obtain the stream function of the spinning cylinder, Ψ, by adding the stream function of a vortex, ψ_{vortex}, to ψ_C, where the nucleus has been replaced by a cylinder, so that:

$$\Psi = \psi_C + \psi_{vortex}$$

Or

$$\Psi = \left(\frac{\mu}{2\pi r}\sin\theta - U_\infty . y\right) - \frac{\Gamma}{2\pi}\ln\frac{r}{r_0}$$

where the new term ψ_{vortex} is given by:

$$\psi_{voetex} = -\frac{\Gamma}{2\pi}\ln\frac{r}{r_0}$$

Γ is the strength of the vortex
r is the distance of any point from the origin and
r_0 is the radius of the cylinder created around the doublet

By setting $\psi_C = 0$, and at $r = a$, we get an expression for a or the radius of the circle that encloses the doublet, based on the doublet strength, as:

$$a = \sqrt{\frac{\mu}{2\pi U_\infty}}$$

For ease of differentiation, re-writing in polar coordinate and re-arranging, we get:

$$\Psi = U_{\infty r}\sin\theta\left[\frac{\mu}{2\pi U_\infty r^2}-1\right] - \frac{\Gamma}{2\pi}\ln\frac{r}{r_0}$$

Thereafter, the normal and tangential velocity components can be obtained by differentiating ψ using:

$$u_n = \frac{1}{r}\frac{\partial\psi}{\partial\theta} \text{ and } u_t = -\frac{\partial\psi}{\partial r}$$

Since the normal velocity component is zero on the cylinder surface, the total velocity is equal to the tangential velocity at any point on the cylinder surface, and is given by:

$$u_t = 2U_\infty\sin\theta + \frac{\Gamma}{2\pi a}$$

At the stagnation point, the total velocity is zero, so that: $u_t = 0$, giving:

$$\Gamma = 4\pi a \sin\theta$$

or,

$$\theta = \sin^{-1}\left(-\frac{\Gamma}{4\pi a}\right)$$

The above expression can be used to study the movement of stagnation points as the circulation value increases from zero to higher values.

3.2.4.1 Stagnation Point Movement on a Rotating Body (Circular Cylinder)

Four cases of movement of stagnation points with changes in the value of circulation are sketched in Figure 3.9 (a)–(d).

From Figure 3.9 (a), when the circulation is zero, i.e., $\Gamma = 0$, signifying that the cylinder is not rotating, the two stagnation points on the cylinder are located furthest apart, one at the front and the other at the rear. This is the case of a stationary cylinder placed in an incompressible and inviscid flow.

As the circulation value increases, but $\Gamma < 4\pi a U_\infty$, the stagnation points start to move and come closer to each other, as seen in Figure 3.9 (b); eventually they collapse to a point when $\Gamma = 4\pi a U_\infty$, at the bottom of the cylinder, which is shown in Figure 3.9 (c). If the circulation is increased further and $\Gamma > 4\pi a U_\infty$, the stagnation points finally detach and move outside the perimeter of the circle.

It is clear, therefore, that for the stagnation points to remain on the circular cylinder surface, the value of circulation must lie between $0 < \Gamma \le 4\pi a U_\infty \sin\theta$. The point where the two stagnation points collapse into a single point on the cylinder surface is when $\Gamma = 4\pi a U_\infty$. This is also the maximum value of the circulation, since the maximum value of $\sin\theta$ is 1. This value is significant because it furnishes the condition from which the maximum value of the lift produced on a fixed body can be determined. But we have to find a way to link the circulation to lift.

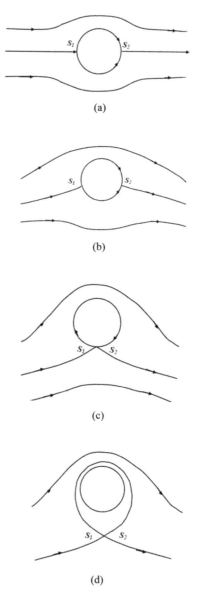

(a)

(b)

(c)

(d)

FIGURE 3.9
Stagnation point movements on a rotating cylinder with changes in circulation value: (a) $\Gamma = 0$; (b) $\Gamma < 4\pi a U$; (c) $\Gamma = 4\pi a U$; (d) $\Gamma > 4\pi a U$.

The connection between circulation and lift force was provided by Zhukovski at the turn of the 20th century. He used the approach of the circulatory streamlines that we discussed earlier. By replacing the nucleus with a rotating cylinder of infinite span and then imparting a translational motion to move it with a given velocity, he deduced that the lift force generated per unit span was equal to the product of the circulation, density, and velocity of the fluid. This is now known as the Zhukovski's theorem of lift, and can be expressed as follows:

$$\frac{L}{S} = \rho_\infty U_\infty \Gamma$$

where the symbols L, S, U_∞, ρ_∞, and Γ denote lift force, span, free stream velocity, free stream density, and circulation, respectively.

It is worth noting that, according to the above expression, lift is independent of the diameter of the circle.

3.2.4.2 Maximum Lift on a Rotating Body (Rotating Circular Cylinder)

Let us now determine the lift coefficient on a rotating cylinder when the stagnation points lie on the cylinder. By definition, the lift coefficient is given by:

$$C_L = \frac{L}{\frac{1}{2}\rho_\infty U_\infty^2 aS}$$

Using the lift theorem of Zhukovski, and replacing the Lift force, L, by $\rho_\infty U_\infty 4\pi a U_\infty \sin\theta S$:

$$C_L = \frac{\rho_\infty U_\infty 4\pi a U_\infty \sin\theta S}{\frac{1}{2}\rho_\infty U_\infty^2 aS} = 4\pi\sin\theta$$

The maximum lift coefficient, when both stagnation points are located on the cylinder, is when sin Θ =1, giving, $C_L = 4\pi$, or approximately 12.56. If a higher lift coefficient is to be achieved, the stagnation points must lie outside the cylinder as shown in Figure 3.9 (d).

3.2.5 Circulation on a Non-Rotating Body (Airfoil)

We have seen how circulation can be created in the previous section. We will now look at creating circulation on a non-rotating body.

In the 19th century, Von Helmholtz [7] discovered that vortices can be used to create circulation around a non-rotating body. But to create these vortices, he suggested two methods. The first method involved using a body with a sharp leading edge inclined to a flow resulting in flow separation at the leading edge of the body. This created a stream of vortices with a dead flow region downstream of it. The second method involved placing a non-sharp leading-edged body in a flow. The friction generated on the body produced a discontinuous flow at the trailing edge and created vortices as the flow left the body.

Kutta [8] and Zhukovski [9], working independently of each other, provided the method to quantify the magnitude of circulation produced on a non-rotating body. Following Helmholtz, they first produced a stream of vortices or a vortex sheet that separated at the trailing edge of a streamlined body with low thickness-to-chord ratio. In the process, Kutta and Zhukovski realized that to produce circulation a smooth separation of the flow from the trailing edge was also needed. This they found was possible when the flow detached at the trailing edge from the upper and lower surfaces with velocities that were equal in magnitude and direction.

While the condition of equal magnitude in velocity could be easily achieved, maintaining the velocity direction so that it was the same at the trailing edge proved difficult. A mathematical solution was possible when both the upper and lower velocities became zero at the trailing edge. This implied that the trailing edge had to be sharp edged and become a stagnation point as shown in Figure 3.10.

Kutta simply enforced a condition of zero velocity to achieve a stagnation point at the trailing edge in his analysis. Zhukovski, on the other hand, produced a more robust mathematical treatment to arrive at the same conclusion.

On a rotating cylinder, as described in Section 3.2.4.2, the maximum circulation was found to be $\Gamma = 4\pi a U$, based on the two stagnation points collapsing onto a single point. This suggested that the movement of the stagnation points could provide a means to determine the maximum lift possible on a fixed body, such as on an airfoil.

Zhukovski [10] demonstrated that this was possible if the method of conformal transformation was applied to transform the rotating cylinder flow into an equivalent flow on a non-rotating airfoil configuration, thereby maintaining flow conformity.

3.2.5.1 Conformation Transformation of a Circular Cylinder to an Airfoil

To apply the conformal transformation method, apart from stream function, we also need its equivalent 90°, out-of-phase function. Ironically, this function is called the "potential function" (note: an ideal flow is also called a potential flow, and so a stream function can be called a potential flow).

A "potential function," too, has "potential lines" similar to "streamlines" of a stream function and can be formed by rotating streamlines through 90°.

FIGURE 3.10
The Kutta–Zhukovski condition of smooth detachment at the trailing edge with $U_{top} = U_{bottom}$.

This makes the potential lines slightly more difficult to visualize to represent the path of flow particles compared to streamlines. The potential functions also have the same units as the stream functions, and their spatial differentiations, therefore, give velocity components.

We can now proceed with the following steps for a two-dimensional flow:

(1). First, the stream function, Ψ, and the corresponding potential function, Φ, of a circular cylinder described in a real plane, Figure 3.11, (x–y plane, coordinates, x, y), are obtained:

(2). Then the cylinder is then represented in terms of a complex function, w, for a complex plane, Figure 3.12, (z-plane, coordinates, x, iy) as:

$$w = \Phi + i\,\Psi$$

(3). The cylinder is transformed into an equivalent airfoil shape in a second complex plane, Figure 3.13, (ς-plane, coordinates, ξ, i), while maintaining geometry and flow conformity.

To achieve the transformation, we need a transformation function to link the two coordinate systems. In other words:

$$\varsigma = f\,(z)$$

There are various transformation functions that can be used. One such function is the Zhukovski function, which is given by:

$$\varsigma = z + \frac{b^2}{z}$$

where b is the Zhukovski constant

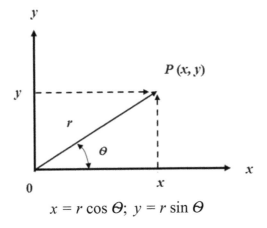

$$x = r \cos \Theta; \; y = r \sin \Theta$$

FIGURE 3.11
The real-plane.

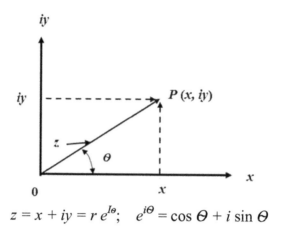

$$z = x + iy = r\,e^{I\theta}; \quad e^{i\theta} = \cos\theta + i\sin\theta$$

FIGURE 3.12
The complex *z*-plane.

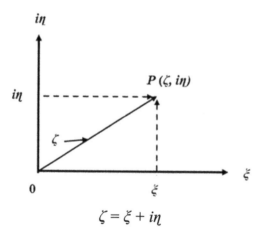

$$\zeta = \xi + i\eta$$

FIGURE 3.13
The complex ζ-plane.

It is worth pointing out that transformation is highly dependent on the values of the parameters: *a* (radius of the circle), *b* (Zhukovski constant), and *r* (distance of a point from the origin of the coordinate system). Thus, when,

$$r = a = b,$$

the circle transforms into a flat plate or,

$$r = a, a > b,$$

the circle transforms into an ellipse.

Similarly, if the following condition is satisfied, the circle will be transformed into a cambered airfoil shape:

$$r \neq a; a \neq b; a = (1+e)$$

It is worth noting that e is the eccentricity that gives thickness to the transformed airfoil. Also it can be shown that the chord length of the airfoil, c, is equal to $4b$.

(4). Once the shapes are obtained, we will impart circulation of equal magnitude to both the cylinder and the airfoil. The movement of stagnation points on the spinning cylinder is related to the flow on the airfoil under the assumption that the flow leaves the trailing edge smoothly, with the trailing edge becoming a stagnation point.

For ease of analysis, the movement of stagnation points on the cylinder, the camber, and thickness of the airfoil will be expressed in terms of angles rather than distances.

We will now describe the process of conformal transformation using Figure 3.14 (a)–(d). We will start by placing a circle in the first complex plane, the z-plane, with the center of the circle offset from both the axes at an angle of incidence of α, as shown in Figure 3.14 (a). The offset is given to produce the thickness and camber in a transformed airfoil shape.

The Zhukovski transformation function is used to establish conformity between the flow field in the z-plane and second complex plane, ζ-plane, as shown in Figure 3.14 (b). Notice that the stagnations points, S_1 and S_2, on the circle are 180° apart, with the flow pattern remaining symmetrical around the cylinder. On the transformed shape, however, we will find the second stagnation point, S_2, to be located on the top surface of the airfoil.

We now introduce circulation in the form of adding vortex motion to the circular cylinder. This has the effect of producing substantial changes to the flow patterns around the cylinder. The flow symmetry is broken and the stagnations points, S_1 and S_2, move closer to one another as can be seen in Figure 3.14 (c) from. S_1 and S_2, to 'a' and 'b' respectively. On the airfoil, however, the stagnation point must be at the trailing edge. So in Figure 3.14 (d), the stagnation point, S_2, now moves and is now located at the trailing edge of the airfoil at point B while there is also a movement of stagnation point, S_1, point to point, A.

3.2.5.2 Relationship between Stagnation Point Movements of Circular Cylinders and Airfoils

On an airfoil, pressure differences on the top and bottom surfaces are produced by the angle of incidence, camber, and thickness of the airfoil. Hence the magnitude of lift needs to be expressed in terms of these two parameters. Let α and β represent the angle of incidence and camber of the airfoil,

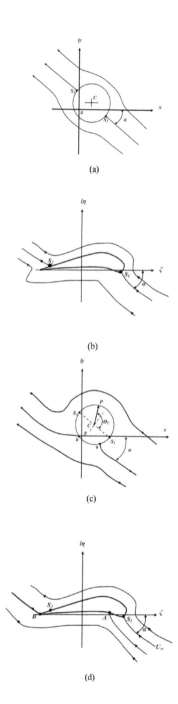

FIGURE 3.14

Conformal transformation: (a) flow around a circle without circulation in the z-plane; (b) flow around an airfoil (transformed) without circulation in the ζ-plane; (c) flow around a circle with circulation in z-plane; (d) flow around an airfoil (transformed) with circulation in ζ-plane.

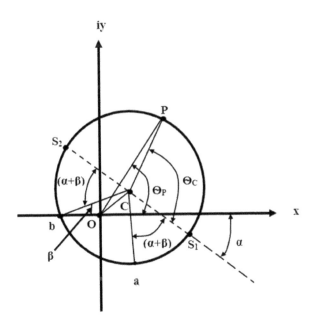

FIGURE 3.15
Geometry to relate circulation to angle of incidence and camber of an airfoil.

respectively. Figure 3.15 has been drawn to illustrate how the circulation on an airfoil can be related in terms of these angles to the value obtained on a rotating cylinder.

On the cylinder, due to circulation, the stagnation point, S_2, has moved to point b because of the Kutta–Zhukovski condition enforcement at the trailing edge. In the process, it has moved through an angle equal to $(\alpha + \beta)$. This also means that the trailing edge stagnation point on an airfoil has moved to point B, as can be seen in Figure 3.14 (d). Replacing θ by $(\alpha + \beta)$, the circulation on the airfoil becomes:

$$\Gamma = 4\pi a \sin(\alpha + \beta),$$

Using,

$$a = b\left(1 + e\right) \text{ and } c = 4b$$

we get:

$$C_L = 2\pi(1+e)\sin(\alpha+\beta)$$

where e can have a theoretical maximum value of 1, giving the maximum value of lift coefficient of an airfoil to be equal to 4π.

In practice, boundary layer growth on the airfoil always gives rise to viscous effects and the flow separates at around 15–18°. Thus, the maximum value of 4π is never achieved. The reported maximum lift coefficient appears to be 1.8 [11], achieved on a NACA 23012 airfoil. Hence various strategies and methods have been deployed to arrest the development of boundary layers and generate the high lift necessary during take-off and landing. We will discuss some of these high-lift devices before considering the Coanda effect as a circulation control device.

3.3 Lift Augmentation

The devices that are deployed for lift augmentation on an airfoil are called flaps. The flaps can either be unpowered or powered. We will discuss both of these types before considering their relevance to the Coanda effect. Before that, we will consider the "thin airfoil theory" to get some basic understanding how we can theoretically determine lift force generated on airfoil. We will then proceed to apply this theory to quantify lift augmentation on an airfoil.

3.3.1 Thin Airfoil Theory

The essence of thin airfoil theory is that an airfoil can be replaced by its mean camber line, and the mean camber line subsequently by point vortices [12, 13], as shown in Figure 3.16. The following notations and assumptions will be used to describe the thin airfoil theory.

NOTATIONS

Cc: chord length
C_L: lift coefficient
A_0, A_1, A_n: coefficients
U_∞:   free stream velocity
θ: angular representation of distance
Γ: circulation
γ: vorticity per unit length
s: distance
ds: elemental distance

3.3.1.1 Assumptions

- The angle of incidence is small, $< 4°$.
- The camber is small.

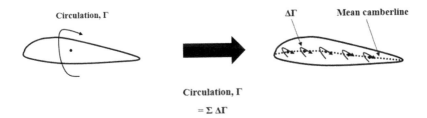

FIGURE 3.16
An airfoil is replaced by its mean camber line.

- The airfoil is thin, and in the limit it does not have any thickness so that it can be represented by its mean camber line.
- The flow is ideal, that is, no viscous effects are present.
- The total circulation is not concentrated as a vortex on the airfoil and the vorticity is distributed to produce the mean camber line, so that:

$$\Gamma = \sum \Delta\Gamma = \sum \gamma(s)\Delta s$$

or

$$\Gamma = \int_0^c \gamma(s)\,ds$$

- $\gamma(s)$ is zero at the trailing edge (Kutta–Zhukovski condition).

- x can be represented in terms of c and θ by $x = \dfrac{c}{2}(1-\cos\theta)$

Since camber is assumed to be small, $\gamma(s)$ also becomes a function of θ. The task in thin airfoil theory then boils down to finding an appropriate expression for the vorticity distribution $\gamma(\theta)$.

Using the assumed relationship,

$$x = \frac{c}{2}(1-\cos\theta)$$

an expression for $\gamma(\theta)$ is obtained so that:

$$\gamma(\theta) = 2U_\infty \left[A_0 \frac{1+\cos\theta}{\sin\theta} + \sum_{n=1}^{\infty} A_n \sin n\theta \right]$$

where,

$$A_0 = \alpha - \frac{1}{\pi}\int_0^\pi \frac{dy}{dx}\,d\theta$$

$$A_n = \frac{2}{\pi} \int_0^\pi \frac{dy}{dx} \cos n\theta \, d\theta$$

Then, for the unflapped case:

$$C_L = \pi \left[2A_0 + A_1 \right]$$

3.3.2 Unpowered Flaps

The conventional high-lift systems generally include mechanical flaps. Mechanical flaps are called unpowered flaps. They form integral part of the wing and are installed at the trailing edge, leading edge, or at both. At the trailing or leading edge, the flaps during take-off or landing may be extended beyond or deflected downward from the main wing.

The choice of the flaps are defined by the mission and performance desired of an aircraft. They generally involve considerations of the lift coefficient, drag coefficient and moment coefficient variations with the angle of incidence (particularly, below the stall angle) as well as flap angle. Figure 3.17 shows a selection of the various two-dimensional representations of mechanical flaps used on airfoils.

3.3.2.1 Thin Airfoil Theory Applied to Unpowered Flaps

The analysis will be carried out using Figure 3.18.

NOTATIONS

c_0:   main chord length
c_F: flap chord length
c: total chord length
α: angle of incidence
α_e: effective angle of incidence
β_F: flap angle
U_∞:   free stream velocity

3.3.2.1.1 Assumptions

- small angle, $\alpha_e = \alpha + \beta_F \dfrac{c_F}{c}$

- constant slope of c_0, $\dfrac{dy}{dx} = \alpha_e - \alpha = \beta_F \dfrac{c_F}{c}$

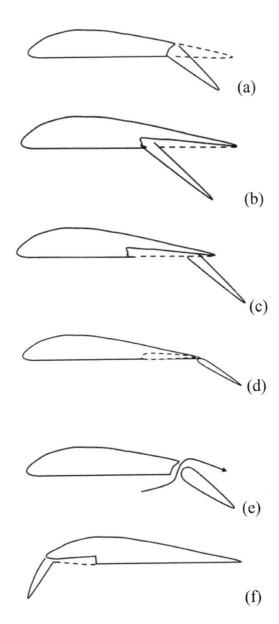

FIGURE 3.17
Sketches of unpowered flaps: (a) plain flap; (b) split flap; (c) Fowler flap; (d) Zap flap; (e) slotted flap; (f) Kruger flap.

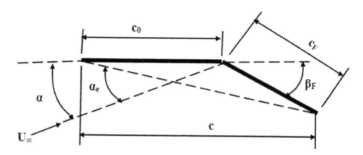

FIGURE 3.18
Geometry of an unpowered flap.

Then, the slope for C_F:

$$\frac{dy}{dx} = \beta_F \left(\frac{c_F}{c} - 1 \right)$$

$$A_0 = \alpha + \beta_F \frac{c_F}{c} - \frac{1}{\pi} \int_0^{\theta_1} \beta_F \frac{c_F}{c} d\theta - \frac{1}{\pi} \int_{\theta_1}^{\pi} \beta_F \left(\frac{c_F}{c} - 1 \right) d\theta$$

or,

$$A_0 = \pi + \beta_F \frac{\pi - \theta_1}{\pi}$$

where, $\theta_1 = \cos^{-1}\left(1 - \frac{2c_0}{c} \right)$

Similarly,

$$A_n = \frac{2\beta_F}{\pi n} \sin(n\theta_1)$$

And substituting in C_L,

$$C_L = 2\pi \left(\alpha + \beta_e \beta_F \right)$$

Where, the flap effectiveness, β_e (change in the effective angle of incidence of the airfoil with changes in the angle of the flap),

$$\beta_e = 1 - \frac{\theta_1 - \sin\theta_1}{\pi}$$

The increase in lift coefficient due to the flap is given by Figure 3.19,

$$\Delta C_L = 2\pi \beta_e \beta_F$$

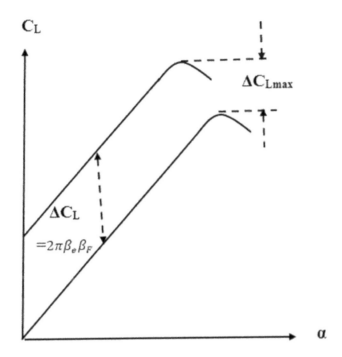

FIGURE 3.19
Lift characteristics of a flapped airfoil.

The increase in C_{Lmax} is lower than the increase in C_L for a single flap, generally one-third lower. More accurate values can be obtained in References [14, 15]. An interesting feature apparent from these references is that at a Reynolds number of 6×10^6 or higher (based on extended cord length), the maximum coefficient of lift is approximately 2.7, irrespective of the type of flap used [16]. Figure 3.20 shows the pressure distribution on a multi-surface airfoil.

The mechanical flaps, however, incur significant weight and volume penalties in wing assembly. These assemblies are also very complex and the flaps perform poorly at off-design conditions. The reduction of weight and simplification of flap designs without compromising performance require special considerations in aircraft design.

It is important to note that this value of 2.7 can be increased to as high as 3.7 if some form of boundary layer control is applied [16, 17].

3.3.2.2 Examples of Unpowered Flaps in Operation

Various flaps have been deployed on many aircrafts over the years. Figures 3.21 through 3.28 show images of some of them.

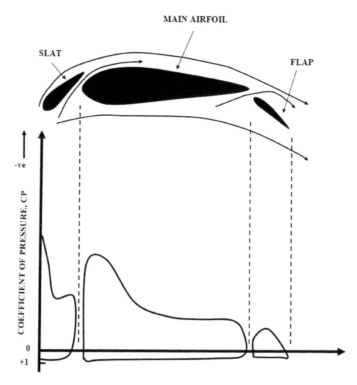

FIGURE 3.20
Pressure distribution on an airfoil with leading edge slat and trailing edge flap.

FIGURE 3.21
North American T-6 trainer showing its split flaps. (After: 2009 [18].)

FIGURE 3.22
The three orange pods are fairings streamlining the flap track mechanisms. The flaps (two on each side, on the Airbus 319) lie directly above these. (After: [19].)

FIGURE 3.23
Plain flap at full deflection. (After: ILA-boy 2008 [20].)

3.3.3 Powered Flaps

Strictly speaking, blown flaps and pure jet flaps should be considered powered flaps. As high-lift device concepts, however, we will include descriptions of reverse flow airfoils and jet airfoils in this section. The cross sections and experimental results regarding performance of some of these powered,

FIGURE 3.24
Split flap on a World War II bomber. (After: "the real KAM75", 2010 [21].)

FIGURE 3.25
Double slotted Fowler flaps extended for landing. (After: AlexHe34, 2011 [22].)

high-lift devices developed at Boeing Aircraft Company in the USA can be found in References [16, 37, 38].

3.3.3.1 Blown Flaps

Blown flaps involve both mechanical flaps and air jets. In this arrangement, air improvement in the effectiveness of mechanical flaps is attempted by installing high-velocity air jets on their upper surface. When only air jets are used and mechanical flaps are dispensed with completely, the arrangement

FIGURE 3.26
Krueger flaps and triple-slotted trailing-edge flaps extended for landing. (After: Pingstone, 2002 [23].)

FIGURE 3.27
Flaps during ground roll after landing, with spoilers up, increasing drag. (After: ShareAlike 4.0 [24].)

is called the "pure jet flap." The pure jet flaps are deflected directly from the trailing edge to produce the same effects as the mechanical flaps.

The concept of the blown jet flap has been around for a long time. Schubauer [26] was probably the first who explored the use of jets to enhance lift. The concept garnered greater attention with the development of turbojets that furnished a readily available means of blowing air. Further impetus came with the realization that the lifting system could be combined with the propulsion system. Consequently, various designs of pneumatic devices that use blown jets have been investigated. Most of them can be considered to be unconventional design concepts with varying degrees of success. We will

FIGURE 3.28
The position of the leading edge slats on an airliner (Airbus A310-300). (After: Pingstone, 2002 [25].)

describe briefly some of these non-conventional designs as they offer interesting insight into fluid motion problems and may provoke innovative ideas in future.

After WWII, workers such as Lighthill [27], Glauert [28], and Williams [29] proposed two-dimensional airfoils that used suction through a slot to produce sink effect to arrest the occurrence of sharp adverse pressure gradients, thereby avoiding or delaying flow separation. These works were, in a sense, a continuation of boundary layer control attempts that had started with Prandtl [30]. The process has been aptly described by Lighthill [31] as follows:

> If the drop from the summit to the foothills is replaced by a shear principle, by a discontinuity in fact where the boundary layer is sucked away, and the remainder of the velocity curve given an even declivity down to the trailing edge, it is hoped that breakaway will be avoided.

Glauert [32] refined his earlier work [28] and produced the interesting finding that it is possible to design thick airfoil shapes over which flow could remain attached and laminar. The qualitative descriptions of the results of his studies in terms of pressure distributions on the top and bottom surfaces

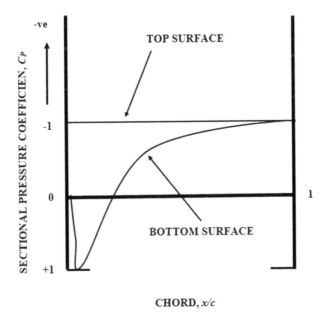

FIGURE 3.29
Qualitative description of the pressure distributions over a lobsterpot airfoil with suction at the trailing edge.

are shown in Figures 3.29 and 3.30. In both cases the pressure on the top surfaces have been maintained constant, and suction has been used to control the boundary layer growth and the adverse pressure gradient. In the first figure, the suction was applied at the trailing edge of a lobsterpot airfoil while in the second figure the suction was applied at a location on the top surface of a thick airfoil. These theoretical trends were validated by the experimental works of Keeble and Atkins [33].

Küchemann [34] observed that the constant pressure shapes and the lobsterpot airfoil studied by Glauert [32] were similar to the cavitation bubbles behind solid bodies as reported by Birkhoff and Zarantonello [35] and Woods [36].

Various high-lift devices employing internally and externally blown flaps and lifting characteristics against blown coefficients can be found in References [37, 38].

3.3.3.2 Reverse Flow Airfoils

Figure 3.31 shows a ducted airfoil that has reverse flow taking place inside. Flow features of such a shape were considered by Küchemann [39]. The air that comes out from its front end bifurcates into two streams that flow around the two parts of the airfoil before joining again at the rear end; then they re-enter the airfoil and repeat the journey continuously.

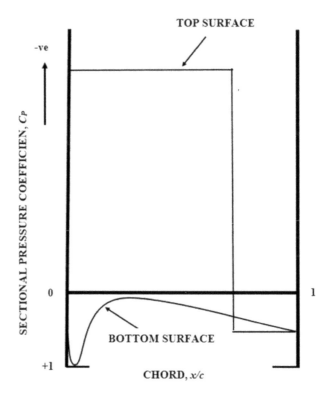

FIGURE 3.30
Qualitative description of the pressure distributions over an airfoil with a suction at the upper surface.

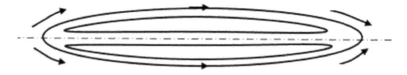

FIGURE 3.31
Schematic of a reverse flow airfoil.

In this type of airfoil, the requirement of the existence of a free stagnation point at intake makes the task of designing the inlet slot a difficult task. The pressure that exists at the stagnation point means that the air has to be pushed in rather than being sucked in, with the consequence that the stagnation point no longer remains fixed and the flow becomes unsteady. Heughan [40] has shown that such unsteadiness can be overcome by placing a solid plate symmetrically along part of the streamline to stop the movement of the stagnation point. Since the flow accelerates on the flat plates, boundary layer growth is inhibited and poses little difficulties.

Using the hodograph method, different shapes of the frontal nose to eject the flow in a manner similar to the reverse flow airfoil concept have been calculated by Eminton [41]. But to keep the flow inside the duct moving continuously with low drag requires energy input. Cost effective means of producing such energy, and delivery of it, are essential to make the concept of the reverse flow airfoil viable.

The fundamental propulsion aspects associated with such airfoils have been considered by Edwards [42] and others. However, to date, it has been difficult to generate a complete lifting body shape with internal duct geometry and a free stagnation point at the rear that produces minimum viscous wake.

3.3.3.3 Jet Airfoils

Figure 3.32 shows a jet airfoil. Similar to the reverse-flow airfoil, this type of airfoil also has a duct passing through it and the flow enters at the front and leaves at the rear without changing direction. In this arrangement, a turbojet engine with by-pass fans, where the by-pass duct is divided into two cold-air ducts on either side of the gas generator may be readily integrated with the wing, thus combining the generation of lift and thrust for the wing. This means the jet wing will require a large number of fans. Keeping the losses and noise produced from these fans to a minimum are fundamental challenges that need to be overcome. Other pneumatic schemes similar to the jet wing concept to augment lift have also been considered with very little success [43–47].

3.3.3.4 Pure Jet Flaps

The schematic of a pure jet flap is shown in Figure 3.33. In this configuration, the jet emerges from a slot slightly upstream of the trailing edge of a small flap and remains attached to it at the exit. It is interesting to note that lift is produced on flapped jet airfoils even if the angle of incidence is zero.

The term jet flap came about from the analogy between a mechanical flap and the jet effects. But, as Küchemann [34] has pointed out, this is only partially justified. The analogy holds in the sense that any vorticity downstream of an airfoil may induce a lift on it, but otherwise the analogy is questionable.

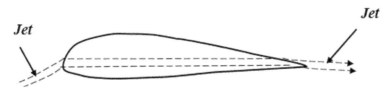

FIGURE 3.32
Schematic of a jet airfoil.

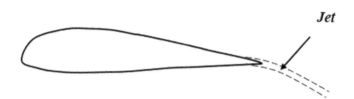

FIGURE 3.33
Schematic of a pure jet flap.

The term "jet flap" continues to be widely used in the literature. We will, therefore, retain the use of the term.

The flow over the jet flap is a demonstration of the Coanda effect [48]. The flow leaves tangentially to the trailing edge. The jet is curved and the angle at which it leaves the trailing edge relative to the mainstream is not constant and can be varied. Across this curved jet, a significant pressure difference is created. This in turn produces a load at the trailing edge. This is highly significant; apart from lift being produced from circulation around the airfoil, there is now the additional circulation from the reaction of the jet.

The jet flap concept as an entity separate from direct studies of Coanda effect has been attempted for a long time. The first known study on jet flaps is attributed to Hagedorn and Ruden [49]. Just as with other schemes requiring jets, the jet flap studies also garnered renewed attention with rapid development of the jet engine after WWII [50–54]. Works by Küchemann [55] have been noteworthy and have helped to provide clarifications to some of the important flow features pertaining to jet flap flows.

The theoretical works of Spence [56, 57] are a significant milestone. Although he used assumptions of small perturbations and linearized two-dimensional thin-airfoil theory, he was able to predict the performance of jet flaps with extremely good effect. Experimental work carried out by Dimmock [58], Malavard et al. [51] and others largely supported his theoretical concepts and results. We will discuss Spence's work in more detail later.

The *Hunting H 126* [59, 60] is the first research aircraft built that used the jet flap concept. It carried out successful flights from March 1963 onwards with lift coefficients of up to 7.5. The aircraft had sharp stalling characteristics requiring a sufficient stall margin. This along with lateral control characteristics limited the maximum usable C_L value to 5.5 [62–64]. Figure 3.34 [61] shows a picture of *H 126*.

3.3.3.5 Thin Airfoil Theory Applied to Powered Flaps

The thin airfoil theory can also be applied to powered flaps. The thin airfoil theory analysis is similar for both blown and pure jet flaps. Sketch of a blown flap geometry is shown in Figure 3.35.

FIGURE 3.34
The H.126's blown flaps with wing tip thrusters and main exhausts. (After: Mark.murphy [64].)

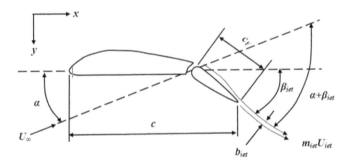

FIGURE 3.35
Sketch of powered flap geometry: blown flap.

There are two models that we will consider; one is the Spence flap model [56, 57], and the other is the Küchemann Flap model [65]. Both are concerned with producing lift through the use of jets. The analysis appears similar, too. The models adopted in the two approaches are, however, open to different interpretations. We, therefore, need to appreciate the approaches of both and adapt our strategies accordingly to best suit the design requirements.

In Figure 3.35, the notations used are:

b_{jet}: thickness of the jet
C:　 total chord length
c_F:　 flap length

m_{jet}: mass flow rate of the jet
U_{jet}: mean jet velocity
U_∞: free stream velocity
A: angle of incidence
β_{jet}: angle with which the jet leaves or the jet flap angle

Let us also consider a segment of a jet as shown in Figure 3.36.

Where
    Δp pressure difference across jet
    R radius of curvature of jet
    U_{jet} velocity of jet
    $\Delta \theta$ differential element of jet

We now apply the momentum theorem to the differential element, $\Delta \theta$, of the jet:

$$m_{jet} \cdot U_{jet} \Delta \theta = \cdot \Delta p \cdot R \cdot \Delta \theta,$$

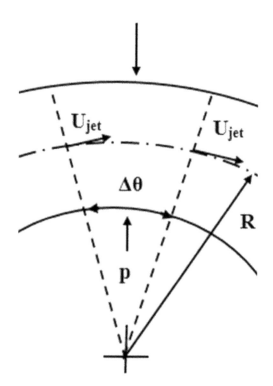

FIGURE 3.36
Geometry of a jet segment.

or,

$$\Delta p = \frac{m_{jet} U_{jet}}{R}$$

3.3.3.6 The Spence Model

3.3.3.6.1 Theoretical Analysis

The notations shown for blown flaps can also be used to explain the Spence flap model that is applicable to both blown and jet flapped airfoils. We will describe Spence's analysis with reference to Figure 3.35 and Figure 3.36.

The standard assumptions of thin airfoil are assumed to be valid. The jet is assumed to be thin, with a finite momentum flux, μ, that acts along the jet, so that:

$$\mu = \rho_{jet} U_{jet} b_{jet}$$

We will further assume that as the jet angle tends toward a zero value. Thus a constant momentum flux acting along the jet gives:

$$\mu = \frac{1}{2} C_\mu \rho_\infty U_\infty^2 c$$

where C_μ is the coefficient of the jet momentum flux.

Equating the terms for μ we get:

$$C_\mu = \frac{\mu}{\frac{1}{2} \rho_\infty U_\infty^2 c}$$

The momentum flux appears to behave like a jet force or a reaction force. In such comparisons, C_μ can be likened to the coefficient of lift. In fact, an approximate value of C_μ is often determined by measuring the reaction statically.

Some entrainment of the surrounding air will inevitably occur. This will introduce errors in the measurement. A more accurate value should instead be obtained by measuring the mass flow rate assuming an isentropic expansion of the jet from the point of its introduction.

Let us now place an infinitesimal number of vortices, each of strength, say, γ_{jet}, along the jet such that the reaction produced on each element is equal to F_{jet}:

$$F_{jet} = \rho_{jet} \cdot U_{jet-\gamma_{jet}} \cdot R \cdot \Delta\theta$$

Also, on flow external to it, the jet produces a reaction, F_{jet}

$$F_{jet} = \Delta p R \Delta\theta,$$

Equating the terms for F_{jet}:

$$\gamma_{jet} = \frac{m_{jet} U_{jet}}{\rho_\infty R U_\infty}$$

When the jet is nearly horizontal, the radius of curvature is very large, and taking x as the distance downstream of the airfoil, the radius in the above expression can be approximated by:

$$\frac{1}{R} \approx -\frac{d^2 y}{dx^2}$$

Then:

$$\gamma_{jet} = -\frac{m_{jet} U_{jet}}{\rho_\infty U_\infty} \frac{d^2 y}{dx^2}$$

So that the total circulation due to the jet is given by,

$$\Gamma_{jet} = \int_0^\infty \gamma_{jet} dx$$

$$= -\frac{m_{jet} U_{jet}}{\rho_\infty U_\infty} \int_0^\infty \frac{d^2 y}{dx^2} dx$$

Since, at a small angle, $\dfrac{dy}{dx} = \alpha + \beta$; and at ∞, $\dfrac{dy}{dx} = 0$,

$$\Gamma_{jet} = \frac{m_{jet} U_{jet}}{\rho_\infty U_\infty} (\alpha + \beta)$$

The total lift, L, generated is composed of lift from circulation, L_c, around the airfoil and lift, L_{jet}, from the vertical component of the jet reaction:

$$L = L_c + L_{jet}$$

where

$$L_c = \rho_\infty U_\infty \Gamma_c$$

and

$$L_{jet} = \rho_{jet} U_{jet} \Gamma_{jet}$$

giving

$$C_L = \frac{2\Gamma_c}{U_\infty c} + C_\mu (\alpha + \beta),$$

or,

$$C_L = A(C_\mu)\alpha + B(C_\mu)b_{jet}$$

For given values of the angle of incidence, α, and C_μ, and taking $b_{jet} = \dfrac{dy}{dx}$ at the trailing edge, Spence used thin airfoil theory and Fourier series analysis for a symmetrical airfoil and obtained results for the vorticity distributions on the airfoil and the jet. Without delving into the mathematics further, we can mention the approximate expression for the coefficients, A and B from his analysis that are valid for $0 < C_\mu < 10$:

$$A(C_\mu) = 2\pi + 1.152C_\mu^{0.5} + 1.106\,C_\mu + 0.051C_\mu^{1.5},$$

and

$$B(C_\mu) = 2\pi^{0.5}C_\mu^{0.5} + 0.325\,C_\mu + 0.156C_\mu^{1.5}$$

Following Spence's analysis [57], typical trends for the vorticity distributions along the jet can be obtained, as shown in Figure 3.37.

3.3.3.6.2 *Validation of the Spence Model*

Overall, Spence's method has resulted in predictions that have been found to agree well for flap deflection angles of as high as 60°.

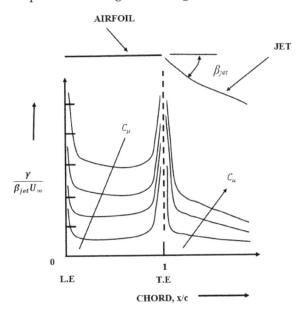

FIGURE 3.37
Vorticity distribution trends along a two-dimensional jet-flapped airfoil.

As lift is produced, it also results in a tangential force component being produced. The magnitude of the coefficient of this tangential force component can be expressed as [34]:

$$C_T = -C_\mu \left(1 - \cos \beta_{jet}\right)$$

There have also been attempts to incorporate the effects of thickness in the expressions for the lift coefficients. These can be expressed as [66]:

$$C_L = \frac{2\Gamma_c}{U_\infty c}\left(1 + \frac{t}{c}\right) + C_\mu\left(\alpha + \beta\right)$$

Maskell and Spence [67] also attempted to develop a theory (for three-dimensional jet-flapped wings) to examine the effect of blowing on induced drag and produced the following relation taking into account thickness, $\left(\frac{t}{c}\right)$, and the aspect ratio, A:

$$C_L = F\left[\left(\frac{\partial C_L}{\partial \beta_{jet}}\right)\beta_{jet} + \left(\frac{\partial C_L}{\partial \alpha}\right)\alpha\right]\left(1 + \frac{t}{c}\right) - \frac{t}{c}C_\mu\left(\alpha + \beta\right)$$

Where

$$\frac{\partial C_L}{\partial \beta_{jet}} = \sqrt{4\pi C_\mu\left(1 + 0.151\sqrt{C_\mu} + 0.139C_\mu\right)}$$

$$\frac{\partial C_L}{\partial \alpha} = 2\pi\left(1 + 0.1514\sqrt{C_\mu} + 0.219C_\mu\right)$$

$$F\left(A, C_\mu\right) \approx \frac{A + 2\dfrac{C_\mu}{\pi}}{A + 2 + 0.604\sqrt{C_\mu} + 0.876C_\mu}$$

Further developments on three-dimensional flapped wing theory have been carried out by Lissamann [68, 69], Lopez and Shen [70], Kerney [71], Shen et al. [72], and others.

From the above works, as has been found for any conventional wing, the lift produced on a jet-flapped wing was also affected by the finite span and was lower due the production of the associated induced drag. The blowing, however, had a favorable effect and produced some reduction in the induced angle. The following expression derived by Spence [57] for induced angle was used, $\alpha_{induced}$:

$$\alpha_{induced} = \frac{C_L}{\pi A + 2C_\mu}$$

This enabled the induced drag coefficient to be determined using $C_{D\,induced}$, so that the total drag coefficient became:

$$C_{D\,induced} = \frac{C_L{}^2}{\pi A + 2C_\mu}$$

$$C_{D\,Total} = C_{D\,viscous} + C_{D\,induced}$$

For attached flow on the wing, the drag due to viscous effects generally remain independent of the angle of incidence and aspect ratio.

3.3.3.7 The Küchemann Jet Flap Model

A case for a simpler jet flap model based on the Thwaites flap has been advanced by Küchemann [65] to provide greater insight into the jet flap principle, as shown in Figure 3.38.

The Thwaites flap is a small, thin flap that is attached at the lower surface and near the trailing edge, but perpendicular to the surface. The trailing edge is different from conventional sharp ending trailing edges in the sense that it is rounded. The Thwaites flap fixes the position of the rear dividing streamline. Boundary layer control through some form of suction, blowing, or entrainment mechanism is needed to ensure that the separated layer at the trailing edge remains thin.

Pankhurst and Thwaites [73] had earlier shown that the Thwaites flap worked well with distributed suction applied to a porous cylinder achieving an impressive lift coefficient of approximately 9. Küchemann argued that a jet introduced at the trailing edge could produce similar effects to those of the Thwaites flap; this could be done by fixing the rear stagnation point beyond the conventional trailing edge stagnation point location, thus increasing the circulation produced.

Küchemann has also shown that his jet flap model can be extended to include other significant effects such as that of thickness, camber, finite span, compressibility, and sweep for design purposes. For example, to ensure uniform loading on the wing chord, generally, camber could be introduced at the leading and trailing edges.

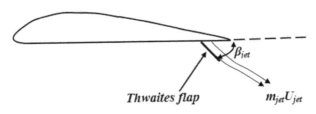

FIGURE 3.38
Schematic of Thwaites flap.

3.3.3.7.1 Accounting for Thickness in Kücheman's Model

The effect of thickness is explained in Küchemann's model. Figure 3.39 shows a qualitative description of the pressure distribution that can be generated using thin airfoil theory and Küchemann's model on a two-dimensional elliptic airfoil.

The pressure distributions predicted higher acceleration both on the top surface as well on the bottom surface compared to thin airfoil predictions. The overall lift force appeared to remain largely unchanged. Experimental results of Dimmock [58] supported the Küchemann model predictions. Despite viscous effects being present, particularly at the jet exit, the results clearly demonstrated that the introduction of the jet provided the benefit of overcoming the adverse pressure gradient built up by energizing the flow.

The effect of thickness on Küchemann's model can also be applied to a three-dimensional flapped wing, and a qualitative comparison is shown in Figure 3.40.

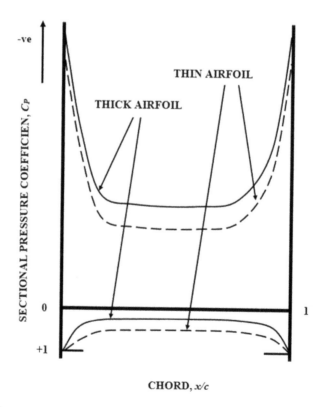

FIGURE 3.39

Qualitative description of the pressure distribution over a two-dimensional flapped elliptic airfoil.

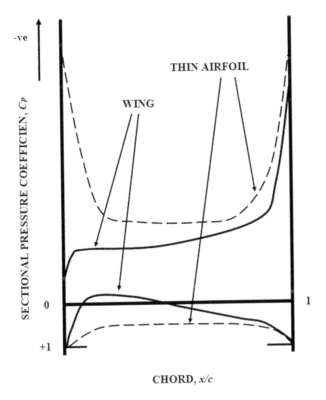

FIGURE 3.40
Qualitative description of the pressure distribution over a three-dimensional flapped wing.

Experimental results of Williams and Alexander [74] on the effects of span provide clear support for the suitability of the Küchemann model, and showed that the model could be used for design purposes. Once again, as was observed for the two-dimensional case, the introduction of jets over a three-dimensional flapped wing provides additional energy to overcome the viscous effects present and mitigates the adverse pressure gradient growth.

3.3.3.7.2 Insight into Flow Physics Using Kucheman's Model

Küchemann's model assumes that the vorticity along the jet can be ignored, and for a given overall circulation, the downwash produced by the loading on the airfoil compensates only the vertical velocity component of the free stream. This ensures that the lift produced from a jet reaction is independent of the angle of incidence, similar to the Spence model.

To reinforce his ideas better, Küchemann [34] considered a two-dimensional flat plate at a zero angle of incidence and introduced a jet from a point near the lower surface but slightly upstream of the trailing edge to fix the stagnation point in a manner similar to the Thwaites flap. This demonstrated fore-and-symmetry of airfoil loading. The downwash velocity becomes zero

from zero upwash and zero downwash. The following qualitative description is based on Küchemann's explanation. Here the loading, $F(x)$, along the
chord is assumed to be made up of two components of loadings, but of equal
and opposite magnitude, one, $F_1(x)$, inducing downwash and the other, $F_2(x)$,
inducing upwash, so that:

$$F \cdot (x) = F_1 \cdot (x) + F_2 \cdot (x)$$

The situation may be likened to that of two flat plate load distributions that
are back to back to each other as can be observed in Figure 3.41.

Insight as to what happens on a three-dimensional jet-flapped wing is also
provided by Küchemann's model where the distinguishing feature of the
flow on a wing is characterized by the shedding of wing tip vortices that
induce a downwash angle on the wing.

For a flat jet-flapped wing placed at zero angle of incidence, the downwash
angle may be assumed to be constant along its chord length. This time, however, the load distributions $F_1(x)$ and $F_2(x)$ are not going to be equal and opposite in magnitude, as was the case for the two-dimensional jet-flapped airfoil.
Therefore, to maintain an overall zero downwash, the loading $F_1(x)$ has to
be reduced accordingly. This would make the chord-wise load distribution

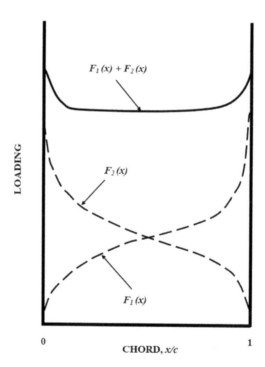

FIGURE 3.41
Qualitative description of the chordwise loading distribution (2D) of a jet-flapped airfoil.

asymmetric. The suction force at the leading edge will be reduced, making it smaller than the suction force at the trailing edge.

The asymmetric aspect in the distribution of bound vorticity and down-wash along the chord of a three-dimensional jet-flapped wing where the front end of the wing is loaded is shown in Figure 3.42. It is possible to restore symmetric distribution by introducing a jet of weak momentum flux on the wing of a very high aspect ratio ($C_\mu \ll \pi A/2$) so that jet-induced thrust becomes equal to the vortex drag.

Küchemann [65] also toyed with the idea that the Thwaites flap be mov-able and thus achieve controlled separation along sharp edges of the nozzle as well as of the flap. This would ensure that the flow is well behaved and efficient.

3.3.3.8 Design Implications of the Spence and Küchemann Models

Both the approaches of Spence [57] and Küchemann [65] can provide useful guidance when it comes to designing jet-flapped airfoils or wings, particu-larly when a designer is confronted with conflicting requirements as how to best achieve the desired outcomes.

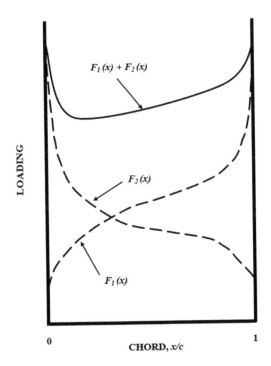

FIGURE 3.42
Qualitative description of the chordwise loading distribution (3D) of a jet-flapped wing.

The two most important decisions a designer probably has to make are concerned with the:

- Shape of the trailing edge, and
- Fixing of the location of the rear stagnation point.

Conventional wings have generally involved a sharp trailing edge to satisfy the Kutta–Zhukovski condition of smooth separation of flow at the trailing edge. However, from a structural and manufacturing consideration, a sharp trailing edge is never the best practical solution. Consequently, a thicker, probably, round trailing shape becomes worthy of consideration.

On a round-shaped trailing edge, the rear stagnation point does not remain fixed, giving rise to unsteadiness with unpredictable behavior of the flow. The introduction of jets can alleviate this problem by fixing the location of the rear stagnation point.

It is important to appreciate the discussion we had before, that to generate high lift, we need to fix the rear stagnation point. The decision of how to fix the stagnation point at a given location will determine the magnitude of circulation we want to produce and hence the magnitude of lift on the airfoil or wing.

We can, in general, adopt two strategies:

- Produce a highly curved jet from the trailing edge itself. This will help maintain a large pressure gradient on the lifting body, or
- Shift the rear stagnation point away from the trailing edge to a point on the lower surface of the airfoil or wing.

The choice of strategy will determine the shape of the trailing edge.

With the first choice, it may involve a sharp trailing edge. If a round-shaped trailing edge were to be used, then the stagnation point must be shifted through progressive deflections of the jet through increments of acute angle steps downstream of the nozzle, similar to the original device that Coanda had experimented with.

For the second choice, a round-shaped trailing edge may be used, but to fix the location of the stagnation, a Thwaites type of flap point may be more suitable.

3.4 Wall Jets

It is clear that the lift augmentation of jet flaps is heavily dependent on the nature of the air jets and how they are introduced. We are interested in a jet that is thin relative to other dimensions in the flow as it flows along a wall surrounded by a fluid which is otherwise at rest or coflowing. We will call this jet a "wall jet" in this section. This jet has higher streamwise velocity than the fluid surrounding it and is likely to be fully turbulent. When the wall jet flows tangentially adhering to the surface, it displays what we have seen before, the features of the Coanda effect.

3.4.1 Straight Wall Jets

For the purpose of easy understanding, we can assume that a two-dimensional wall jet has, on one side, the characteristics of a free jet while on the other side those of a boundary layer. This is, of course, an over-simplification since the wall definitely exerts its influence on the outer region of the flow, and similarly, the outer flow region affects the flow inside of the boundary layer of the wall. The overlap of the two flow regions is determined by the maximum velocity in the jet. The resulting velocity profile will be a composite profile of a half jet and a thin boundary layer, as shown in Figure 3.43.

The earliest experiments on wall jets were performed in 1934 by Foerthmann [75], followed by, in later years, Sigalla [76] and Barke [77].

Foerthmann [75] discovered that the velocity profiles are self-similar. Ignoring the immediate downstream effect from the slit, these profiles could be described by the equation:

$$u \sim x^{-1/2} \cdot f(y / x)$$

where x is a fictitious distance from the exit slit.

Evaluating the distribution of the shearing stresses, Foerthmann determined the mixing length law of the form:

$$y = 0.068 \, b$$

where b denotes the width of the wall jet.

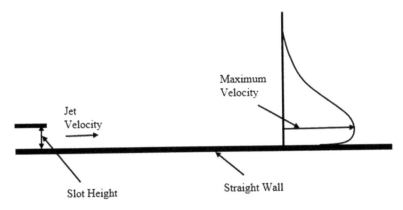

FIGURE 3.43
Velocity profile of a straight wall jet.

The above relationship was also confirmed by the measurements under-taken by Sigalla [76], who established that the local shearing stress can be expressed empirically by:

$$\frac{\tau_0}{\frac{1}{2}\rho U_m^2} = 0.0565\left(\frac{u_m\delta_1}{\upsilon}\right)^{-0.25}$$

where u_m is the maximum velocity of the wall jet, and δ_1 is the corresponding distance from the wall.

For the adiabatic wall temperature for the wall jet, Riley [78] found the recovery factor formed with the maximum velocity can be expressed as:

$$R = 2C_P\left(\frac{T_{adiabatic} - T_\infty}{U_\infty^2}\right) = r(Pr),$$

where the assumptions are:

$$\left(T_{adiabatic} - T_\infty\right) \sim x^{-1}, \text{ and } u \sim x^{-1}$$

For the special case of Prandtl number, $Pr = 1$, $r = 0$, $T_{adiabatic} = T_\infty$, dissipation causes the transport of the total energy produced away from the wall.

We know that any flow over a surface, be it straight or curved, invariably forms a boundary layer. The main difference between normal boundary layers and wall jets is that the wall jets produce a stronger attachment of the jet to surface, hence they are capable of producing the Coanda effect. The wall jets are able to resist the build-up of adverse pressure gradients for a longer distance and consequently delay or prevent flow separation more effectively.

Since no flow can cross the boundary of the wall, this inhibits the growth of pressure reflections from the wall. The wall acts a damping mechanism that restricts the size of the large eddies that may be formed in the outer flow regions. This reduces the entrainment of a wall jet compared to a free jet with subsequent lowering of the level of turbulent energy transfer from the streamwise to the flow in the normal direction.

The above scenario translates into positive shear stress in the outer region and negative shear stress in the inner boundary layer of the wall jet. Compared to a free jet flow, for the wall jet, this has the effect of lowering the Reynolds stress normal to the wall [79]. The point of zero stress, which is generally assumed to coincide with the location of the zero mean velocity gradient, now moves away from it and resides closer to the wall. The assumption of coincidence of the two locations, which is the basis of the simple algebraic turbulence model, no longer remains valid and cannot be applied. Although the logarithmic velocity profile as applicable to the inner layer of the boundary layer may still be valid, this is only the case for a small part of the near-wall flow.

In an adverse pressure gradient, self-similarity laws can be applied to straight wall jets with stream external to it providing the case for non-dimensional properties to be taken to be independent of downstream locations.

Such assumptions, although not strictly true, nevertheless serve as good approximations. Furthermore, the assumptions are valid only if the external velocity to jet maximum velocity ratio remains more or less constant [80]. Overall, flow interactions between the outer and inner layers are complex issues and are beyond the scope of this book.

3.4.2 Curved Wall Jets

When the surface is mildly curved, as happens in the case of blown flaps, the jet thickness is approximately 1% of the radius curvature of the streamwise wall radius. The function of blowing in this case is to maintain flow attachment and delay flow separation. The trailing edge is still sharp, and we can still enforce the usual Kutta–Zhukovski condition on these airfoils.

In such flows, the similarity assumption is valid [81], but only partially, provided the ratio of radius of curvature to the jet width remains constant. The streamwise curvature encountered generates additional strain on the turbulent flow. The surface curvature effects on turbulent flows become quite substantial and difficult to predict accurately. Figure 3.44 shows the curved wall jet and possible shape of the velocity profile of the jet on the wall.

As the curvature of the wall increases, such as on airfoils that have a rounded trailing edge, the jet thickness can increase significantly and become several times larger than that for blown flaps. Since the trailing is no longer sharp, we cannot enforce the Kutta–Zhukovski condition on such configurations. The stagnation point moves outside of the trailing edge, leading to increases in the circulation value and higher lift.

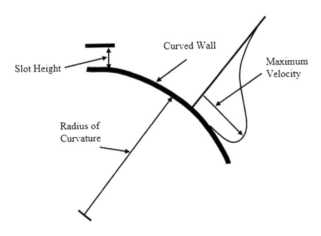

FIGURE 3.44
Velocity profile of a curved wall jet.

Wall jets have been employed in practice for boundary layer control and in film cooling [82]. The attachment of flow length or the movement away of the stagnation point is a function of jet blowing. The amount of jet blowing can be controlled and be used as a circulation control parameter.

The first attempts to describe the circumstances of a wall jet theory was undertaken by Glauert [83]. His work was considerably improved by Eichelbrenner and Dumarge [84]. The semi-empirical theory succeeded for the first time in predicting the separation of a wall jet. Subsequently, based on experimental measurements [85–87], Gartshore and Newman [85] were able to propose an integral momentum method for wall jets with injection and determined the numerical value of the momentum coefficient to prevent the separation of the wall jet.

Experimental and theoretical investigations into the pattern created by two-dimensional plane jets flowing along the contour of a circular cylinder have been carried out by Gersten [88]. Dovarak [89] also performed calculation on two-dimensional turbulent boundary layers on highly convex, curved walls with particular attention to wall jets flowing along curved walls. Sforza and Herbst [90] have investigated three-dimensional aspects in incompressible turbulent walls.

Most of the research on curved wall jets have used logarithmic spirals or circular arcs, mainly because the assumption of self-similarity can be used. From an engineering standpoint, however, the spiral shape is not very useful, and efforts are continuing to develop numerical schemes that would be capable of capturing the curvature effects on any arbitrary shape.

From a physical experimentation point of view, the high intensities present in wall jets flows also present difficulties, and accurate measurements are difficult to obtain. Most researchers have, therefore, attempted to create two-dimensional flows in their studies. A review conducted by Launder and Rodi [79], however, have found the results from these works to be deficient in achieving two-dimensionality.

There appears to be three major reasons for flow non-uniformity in wall jet experiments. One emanates from manufacturing errors in producing test models with the right slot configurations. Fekete [91] found the manufacturing errors to be responsible for wide variations in jet thickness as well as unreliable pressure measurements.

Guitton and Newman [92] and Gartshore and Hawaleshka [93] have found the vorticity created at the end of a slot to be another major cause that produces non-uniformity and three-dimensional effects. They used end plates along the sides of the wall jet to arrest the transverse movement. This, however, gave rise to secondary flows from the interaction between the end wall boundary layer and the wall jet, and created a somewhat U-shaped secondary flow that bifurcated around the end plates. Additional end plates [92] were used to reduce boundary layer growth and reduce secondary flow from developing. Attempts to inject energy into the wall boundary layer through tangential blowing have not been very successful [92].

A concave curvature may be another cause of a wall jet to contract and produce flow non-uniformity. The contraction may lead to the formation of longitudinal vorticity and the spanwise spread of the flow [94]. Large aspect ratio slots are often used to reduce the three-dimensional effects, but considerations of machining such large slots accurately limit their practical applications.

There have been attempts to understand the three-dimensionality effects through studies on free jets, wall jets, and wakes [95–97] from rectangular slots. By studying the decay of maximum velocity, Sforza and Herbst [90, 90] observed three distinct regions:

- Potential core.
- Characteristic decay.
- Axisymmetric decay.

In wall jet studies, although three similar regions can be observed [81], the decay of the maximum velocity was found to be much more dependent on the orifice configuration and its aspect ratio, with the spanwise spread being more rapid. Results from annular wall jet studies of streamwise effects and curvature effects also showed similar trends.

Three-dimensional wall jets with a finite ratio of the two sides have been studied experimentally by several workers [90, 98–100]. The measurements from these studies show a rapid rate of jet spreading in the spanwise direction and the existence of a very different fictitious origin for the growth of the width of the jet in the parallel as opposed to the normal wall direction.

3.5 Circulation Control Airfoils

We are now in a position to talk about circulation control airfoils that utilize the Coanda effect. We have seen earlier in Figures 3.9 (a)–(d) that circulation can be affected by the movement of stagnation points. High lift can be achieved by moving the location of the stagnation point beyond the sharp trailing edge of a conventional airfoil or blunt trailing edge profile. This may be achieved by either suction or blowing.

Prandtl [101] used steady suction to remove the low momentum fluid from the boundary layer from a given free stream on one side of a circular cylinder. This helped to maintain flow attachment on the side of the suction slot to produce lift. The concept has been applied to thick airfoil shapes as well, to prevent or delay flow separation. However, to move the stagnation point away from the trailing edge location without flow separation requires the application of very large suction. This would incur a large drag penalty and lower the aerodynamic efficiency. Moreover, large internal ducts would be required to implement the suction system. These are major drawbacks that have proved very difficult to overcome.

Thus, from practical and performance considerations, the suction methods on their own are not considered suitable for circulation control application and their use has been limited to the prevention or delay of flow separation on an airfoil surface.

Blowing is the other technique that has been found to prevent or delay flow separation on an airfoil surface. Blowing can be used to produce wall jets, which we have discussed in the previous section. Wall jets can, therefore, be considered as another means of controlling the boundary layer. In that sense, Küchemann's [65] concept of blowing over a moveable flap to prevent flow separation and fix the stagnation point at the trailing edge essentially becomes a demonstration of boundary layer control by wall jets. In this scenario, the increment in lift, ΔC_L is roughly proportional to the jet momentum C_μ (i.e., $\Delta C_L \propto C_\mu$) [102].

If the jet-flap effects described in Section 3.3.3 is exploited further and incorporated in Küchemann's procedure, then the stagnation points could be shifted beyond the conventional sharp trailing edge location to any desired location. This movement of stagnation points (Section 3.2.4.1) will have the effect of generating additional circulation, or produce "Super circulation" and, therefore, higher lift. In this regime, the lift increment, however, is proportional to the square root of the jet momentum, i.e., $\Delta C_L \propto \sqrt{C_\mu}$ [102]. The process of creating high lift in this controlled manner forms the basis of a circulation control airfoil.

The review paper by Willie and Fernholz [48] brought the case of the Coanda effect to the fore to complement the circulation control efforts using wall jets. The introduction of the Coanda effect meant additional complexity into circulation control studies because the effect of air entrainment into the jet had to be taken into account. This is highlighted in the studies by Wygnanski [103], who has shown entrainment gives rise to some thrust losses that should not be ignored. Loth [104], on the other hand, has shown that the introduction of boundary layer suction just upstream of the circulation control blowing slot produces entrainment that can have a beneficial effect by increasing the blowing efficiency appreciably.

3.5.1 Lift Enhancement

Early attempts to use circulation control were concentrated toward producing a stoppable rotor VTOL aircraft at the British National Gas Turbine establishment in the mid-1960s. The concept developed by Cheeseman and his colleagues [105, 106] involved a blown two-bladed rotor that would produce high lift during hover, but could also be stopped and stowed within the fuselage of the helicopter during forward flight. To achieve the maximum lift coefficient of 4π on a fixed wing (Section 3.2.4.2), an airfoil with a thickness to chord ratio of unity would be required. A circular cross-sectioned airfoil appears to be the shape that meets this requirement, and in theory, the lift coefficient of 4π can be achieved if the flow and the stagnation points remain attached to the surface, in a manner as shown in Figure 3.9 (c). Further increases in the lift coefficient are possible, such as if the stagnation points detach and lie outside the circular cylinder surface, as is shown earlier in Figure 3.9 (d).

Lockwood [107] was able to achieve coefficient of lift values in excess of 4π by tangential blowing from surface slots on a circular lifting surface. The ability to produce high lift independent of the angle of attack is a major feature of circulation control airfoils as shown in Figure 3.45.

But such high lift production requires high blowing (C_μ), which is also accompanied by the generation of high drag, thus reducing the aerodynamic efficiency of the lifting body to unacceptably low values [108]. Efforts to reduce drag by replacing the circular cylinder with thinner elliptic shape, however, have yielded limited success [109, 110].

The experimental investigations of Englar [111] performed with high velocity Coanda wall jets on bluff trailing edge circulation control airfoils have shown some clear relationship between blowing and slot height. The

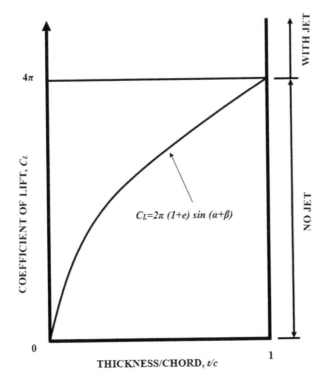

FIGURE 3.45
Qualitative description of the maximum lift coefficient versus thickness-to-chord ratio with and without jet effect.

geometry of a two-dimensional semi-ellipse model was used in this experiment. Greater turning was achieved with a smaller height for a given C_μ. This was attributed to larger flow entrainment as the ratio $\dfrac{V_j}{V_\infty}$ becomes larger for a given C_μ for a smaller slot height. The static pressure distribution results further confirm that high lift can be produced without flaps. The findings are summarized in Figures 3.46 (a) and (b).

3.5.2 Drag Reduction

Drag reduction was key in more advanced circulation control airfoil designs. Englar advocated two approaches [112–114]. The first approach involved a fixed, simple radius reduction while the second included a simple circulation control wing flap.

3.5.2.1 Fixed Radius Reduction

The first approach led to a super-critical type of shape. This airfoil was developed from a low-speed consideration. A summary of the performance in terms of drag polar is shown in Figure 3.47. The main feature evident in this plot is that a slight blowing can reduce C_d and increase C_l at the same time. Details of the experiment and interpretation of the data can be found in Reference 112.

3.5.2.2 Circulation Control Flap Addition

In the second approach, the circulation control airfoil flap had a short chord with a curved upper surface and a sharp trailing edge. The airfoil was dual radius and its leading edge had an inverted tangential slot to replace any mechanical flap. The trend exhibited can be observed in the drag polar plot as shown in Figure 3.48. The results demonstrate the ability to dramatically interchange lift and drag when the flap is deployed. The thrust/drag interchange in this figure was also indicative of high lift and drag for the STOL approach. More details can be found in References 113, 114.

3.5.2.3 Wing Tip Vortex Attenuation

Apart from decreasing aerodynamic efficiency, the trailing vortex of an aircraft poses significant hazards to other following aircraft. The crash of the American Airlines Flight 587 on November 12, 2001 over New York that killed 265 people is attributed to wake turbulence shearing off the tail fin of the Airbus A 300 [115]. Depending on the flight trajectory, several types of induced aerodynamic loadings may be imposed on a following aircraft [116]. Figure 3.49 shows a schematic of an aircraft and its wake where such vortex encounters may become hazardous.

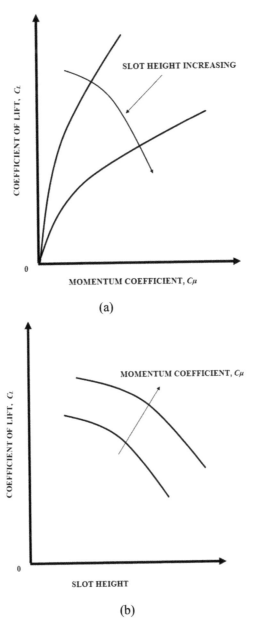

(a)

(b)

FIGURE 3.46
(a) and (b): Qualitative description of the performance of a semi-ellipse circulation control airfoil under varying jet momentums and slot heights.

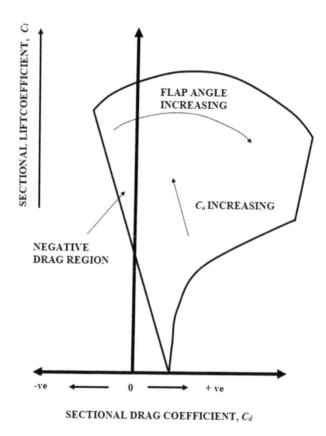

FIGURE 3.47
Qualitative description of the drag polar of a fixed radius reduction circulation control wing.

Simpson et al. [117–119] have studied the vortex-induced wake and the effect of Coanda jets from lift enhancement, drag reduction, and safety considerations. Simpson et al. [119] also explored the potential of Coanda effect as a means of reducing drag through the alleviation of wing tip vortices. They [119] conducted wind tunnel experiments on a NACA 0015 rectangular half wing using straight jets and Coanda jets on both unflapped and flapped wing configurations. The geometry of the test model is given in Figure 3.50.

Vorticity distributions within the vortex wake obtained with a blowing coefficient of 0.065 at a fixed angle of incidence of 10° are given in Figures 3.51 (a) and (b) for the basic wing and flapped case.

3.5.3 Integrated Propulsion and Lift System

The advent and development of jet propulsion were viewed with great excitement as this method offered the prospect of integrating propulsion with lift generation, which has been an important goal of aerospace researchers for

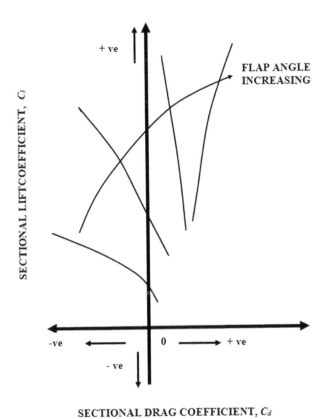

FIGURE 3.48
Qualitative description of the performance of a dual radius circulation control airfoil.

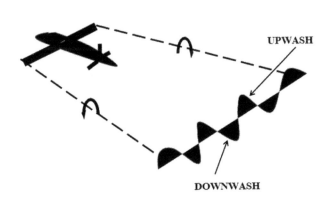

FIGURE 3.49
Vortex wake showing downwash and upwash that may produce hazardous encounters to following aircraft.

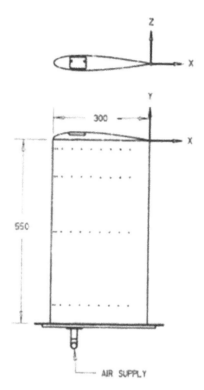

FIGURE 3.50
Geometry of the test model and coordinate system used. (After: Simpson, Ahmed, and Archer, 2002 [119].)

a long time [120, 121]. The interest continues unabated to this day. A simplified block diagram of an integrated wing and propulsion system is given in Figure 3.52.

In the scheme shown in Figure 3.52, the thrust is produced by the primary propulsion system. This thrust is boosted by the reactionary forces of the circulatory system. But since air is bled from the engine to the circulatory control system, the overall thrust-producing capability of the engine is reduced. The objective is to recover the thrust. How much this thrust can be recovered depends on a large number of factors, such as on the losses incurred in the ducting system and the efficiencies of the Coanda nozzle and overall circulation control systems employed.

3.5.3.1 Power Requirements

A critical aspect of the successful development of propulsion integrated lift systems is their total power requirements. Jones [122] has examined the power requirements of typical Coanda jets that could be used in the

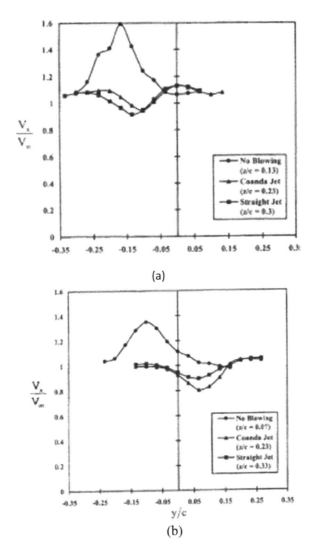

FIGURE 3.51
Vorticity distribution within the vortex wake at downstream location of $x/c = 1.67$. (a) basic wing (unflapped); (b) flapped wing. (After: Simpson, Ahmed, and Archer, 2002 [119].)

integrated propulsion and lift system. He defined C_μ or the momentum or thrust coefficient for two-dimensional flow at the jet exit by:

$$C_\mu = \frac{\text{thrust}}{\frac{1}{2}\rho_\infty U_\infty^2 bc} = \frac{mU_{\text{jet}}}{\frac{1}{2}\rho_\infty U_\infty^2 bc} = \frac{2hw}{cb}\frac{\rho_{\text{jet}}U_{\text{jet}}^2}{\rho_\infty U_\infty^2}$$

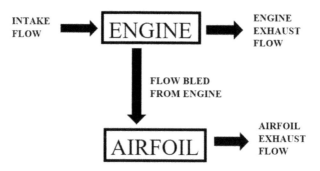

FIGURE 3.52
Block diagram for an integrated wing and propulsion system.

where,

b, c	span and chord of wing
h, w	slot height and width
m	mass flow rate
U_∞, U_{jet}	free stream and jet velocity
ρ_∞, ρ_{jet}	free stream and jet density

Assuming that the power expended was equal to the power necessary for the jet velocity head and the power lost at intake, Jones then went on to obtain an approximate expression for the ideal, non-dimensional coefficient of total fluid power, C_{FP}, as:

$$C_{FP} = \frac{C_\mu^{\frac{3}{2}}}{2\sqrt{2\frac{h}{c}}}\left[1+\frac{4\frac{h}{c}}{C_\mu}\right]$$

The interrelation between the three parameters in the above expression can be graphically observed in Figure 3.53, where the trend is similar to that observed for the lift coefficient in Figure 3.46 (a) and (b).

3.5.3.2 Current Limitations

Finding suitable light-weight and cheap materials that could withstand the heat has been difficult and impeded the development of lift-propulsive integrated systems. As Cerchie et al. [123] have pointed out, with the exception of MIG 21, most production aircraft such as the Lockheed F104 Starfighter, Blackburn NA 39 Buccaneer, and Dassault Etandard IVM, therefore, generate lift using compressed air generated before combustion by engines and ducting the air to slots and blowing it over flaps.

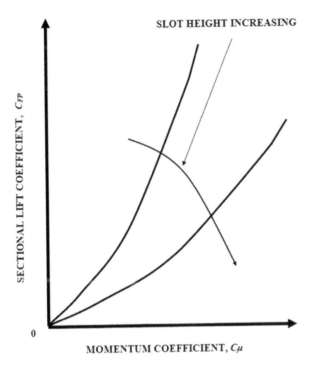

SLOT HEIGHT INCREASING

FIGURE 3.53
Qualitative description of power requirements for Coanda jets with different jet exit slot heights.

3.6 Circulation Control Aircraft

3.6.1 X-Wing Aircraft

An extension of Cheeseman's concept [105, 106] mentioned earlier formed the basis of the X-wing aircraft at the U.S. Navy's David W. Taylor Naval Ship Research and Development Center [124]. In this type of aircraft, the plan was to use circulation control airfoils to produce lift in either direction and allow the rotor/wings to maintain lift and control during starting and stopping rotation [125, 126]. Figure 3.54 shows the diagonal view of Sikorski X-wing.

Converting the X-wing concept into reality was a technically challenging undertaking. The blades had to be double-ended and very stiff to ensure that they could operate in both forward-swept wing and stopped rotor flight modes. The rotor blades had to carry lift without the use of centrifugal force of a conventional helicopter. The use of the Coanda effect to control blade lift required the design of a large compressor and highly complex control systems to feed circulation air to the rotors. The starting and stopping sequence required a high energy clutch and a braking system, and advanced

FIGURE 3.54
Sikorski X-wing diagonal view. (After: NASA, [127].)

computation capability to control the process. All these difficult demands, constant delays to provide adequate solutions, and high cost eventually led to the termination of the project [128].

3.6.2 No Tail Rotor (NOTAR) Flight Vehicle

The term NOTAR is now widely accepted as the description of a helicopter that does not use a tail rotor. The MD-Explorer NOTAR helicopter [129] used by the German police force and shown in Figure 3.55, exemplifies the successful commercial application of the Coanda jet.

FIGURE 3.55
MD Explorer used by the German police. (After: Lehle, 2005 [129].)

The original concept of NOTAR design was conceived in 1975 at Hughes Helicopters Incorporated. A detailed description of the history and development of the NOTAR can be found in Ref. [130].

In the NOTAR design, the conventional tail rotor is replaced by Coanda jets blowing around the tail boom. The design includes an enclosed fan, vertical fan, circulation control Coanda tail boom, and an array of valved turning vanes. The downwash from the main rotor is combined with the Coanda jet that exists tangentially from the circular tail boom. This creates an anti-torque side force on the tail boom. The side force thus created by the circulation control boom provides the majority of the trim thereby counteracting the main torque during hover. Helicopter flight tests have shown that the NOTAR system provides improved safety, as well as having high handling qualities at low speed, good hover stability out of ground effect, and reduced vibration and noise [131].

3.6.3 Coanda Effect UAV/MAV (Unmanned Aerial Vehicle/Micro-Aerial Vehicle)

The term UAV stands for 'unmanned aerial vehicle' or an aerial vehicle that can be operated without any crew. UAVs are often called 'drones'. An MAV or micro-aerial vehicle is a miniaturised version of a UAV. UAV or MAV forms one the components of an overall unmanned aerial system (UAS).

Historically, the earliest versions of UAVs were balloons and unpowered. Nicola Tesla introduced the powered concept of UAVs in 1915 but the successful implementation of powered UAVs occurred in 1991 during the Gulf War when over 300 UAV combat missions were conducted by the US and allied Forces [132].

Since then the role of UAVs has grown in significance, particularly, for military use. This is spelt out in the road map prepared by the US army to use UASs as their eyes [132]. Other military powers also view UASs with similar importance. The commercial and entertainment applications of UAVs are also showing promise and receiving greater attention.

The traditional UAVs differ from Coanda effect UAVs in that the traditional UAVs have fixed wings. These traditional UAVs require runways for take-off and landing. They also have lower hover capabilities. Consequently, rotors similar to those used in helicopters or VTOL/STOL aircraft are often used in the conventional UAVs to improve their manoeuvrability.

The main aims of Coanda effect UAV or MAVs are to produce efficient cruise as well as hover capabilities. The configurations of Coanda effect UAV or MAVs are, therefore, generally semi-spherical or saucer types to help produce Coanda surfaces [133–135].

Coanda effect MAVs have been studied by Ahmed et al. [134] who used saucer type configurations. The authors have developed the relevant equations from conservation principles and obtained the aerodynamics and flight mechanics on their Coanda effect MAV shapes. The analyses, however, were two-dimensional in nature and based on simplifying assumptions. They nevertheless, were able to model the forces and moments on their chosen geometry. Other studies, such as those by Lee et al. [135] considered the stability and control aspects of UAVs. They used a dynamic simulation model where they incorporated approximate servo mappings to achieve the required control on a saucer type Coanda effect UAV.

Most of these studies on Coanda effect UAVs, however, are at their nascent stage and limited to numerical investigations or simulations. Greater efforts are required before they are capable of producing practical solutions for military or commercial applications.

3.7 Concluding Remarks

Technological advancement in new materials and computing power are making the goal of fully integrating lift and propulsion capabilities more attainable. Future efforts are likely to benefit from the introduction of new modes of blowing and methods of flight control. Techniques to manipulate circulation and incorporate the Coanda effect to produce high lift during take-off, low drag during cruising for better fuel economy but high drag during landing requiring shorter runway length, as well as lighter aircraft and lower noise will drive the research of tomorrow.

References

1. https://patents.google.com/patent/US2108652A/en
2. Arnodt, R., HENRI COANDA SS LENTICULAR FLUGSCHEIBE, in Henri Coanda Lenticular Disc – GREY FALCON, http://discaircraft.greyfalcon.us/Coanda.htm
3. Bzuk at English Wikipedia, https://en.wikipedia.org/wiki/Avro_Coanda_VZ-9_Avrocar#media/File:John_Frost_color.jpg
4. Bzuk at English Wikipedia, https://commons.wikipedia.org/wiki/Avrocar_at _factory.jpg
5. Avrocar: Saucer Secrets from the Past, Winnipeg: MidCanada Entertainment. 2002. Archived from the original on 4 August 2006, https://en.wkipedia.org/wiki/Avro_Coanda_VZ-9_Avrocar
6. Rich, P., McKinley, R.J., and Jones, G.S., (2005), Circulation Control in NASA's Vehicle Systems Program, NASA/CP-2005-213509/PT1, pp. 1–36.
7. von Helmholtz, H., (1858) (in German), Über Integrale der hydrodynamischen, Gleichungen, *Journal für die reine und angewandte Mathematik* vol. 23, pp. 215–228.
8. Kutta, M.W., (1902) (in German), Auftriebskraft in strömendon flüssigkeiten, *Illustrierte Aeronautische Mitteilungen*, vol. 6, pp. 133–135.

9. Zhukovski, N.E., (1907), On the adjunct vortices, *Transactions of the Physical Section*, vol. 13, pp. 12–25 (London).
10. Zhukovski, N.E., (1910) (in German), Über die Konturen der Tragflächen der Drachenflieger, *Zeitschrift für Flugtechnik und Motorluftschiffahrt*, vol. 1, pp. 281–284 and (1912) vol. 3, pp. 81–86.
11. McCormick, B.W., (1997), *Aerodynamics, Aeronautics, and Flight Mechanics*, 2nd edition, John Wiley and Sons.
12. Glauert, H., (1924), A theory of thin airfoils, British Aeronautical Research Committee, Report and Memorandum 910.
13. Clancy, L.J., (1979), *Aerodynamics*, Pitman.
14. Abbott, I.H., and von Doenhoff, A.E., (1956), *Theory of Wing Sections*, Dover.
15. Roshko, A., (1959), Computation of the increment of maximum lift due to flaps, Douglas Aircraft, Report SM-23626.
16. Kuethe, A.M., and Chow, C-Y, (1998), *Foundations of Aerodynamics – Basis of Aerodynamic Design*, 5th edition, John Wiley and Sons.
17. Goradia, S.H., and Colwell, G.T., (1975), Analysis of High lift wing systems, *Aeronautical Quarterly*, vol. 26, no. 2, pp. 88–108.
18. By Pline, Public Domain, https://commons.wikimedia.org/w/index.php?cur id=5942654
19. Photograph by Adrian Pingstone, Public Domain, https://commons.wikimedia .org/w/index.php?curid=10902915
20. By ILA-boy, Public Domain, https://commons.wikimedia.org/w/index.ph p?curid=4216676
21. By "the real Kam75", Public Domain, https://commons.wikimedia.org/w/ind ex.php?curid=25762940
22. By AlexHe34, Public Domain https://commons.wikimedia.org/w/index.ph p?curid=12727420
23. Photograph by Adrian Pingstone, Public Domain, https://commons.wikimedia .org/w/index.php?curid=118310
24. https://creativecommons.org/licenses/by-sa/4.0/deed.en
25. Photograpgh by Adrian Pingstone, Public Domain, https://commons.wiki media.org/wiki/File:Wing.slat.600pix.jpg
26. Schubauer, G.B., (1933), Jet propulsion with special reference to thrust augmentation, NACA TN 442.
27. Lighthill, M.J., (1945), A new method of two-dimensional aerodynamic design, ARC R & M 2112.
28. Glauert, M.B., (1945), The design of suction airfoils with very large CL range, ARC R & M 2111.
29. Williams, J., (1950), Some investigations of thin nose-suction aerofoils, ARC R & M 2693.
30. Prandtl, L. (1904) (in German), Über Flüssigkeitsbewegung bei sehr kleiner Reibung, in *Proceedings of 3rd International Mathematical Congress*, Heidelberg, Germany.
31. Lighthill, M.J., (1945), A theoretical discussion of wings with leading edge suction, ARC R & M 2162.
32. Glauert, M.B., (1947), The Application of the exact method of airfoil design', ARC R & M 2683.
33. Keeble, T.S., and Atkins, P.B., (1951), Tests of Willium Glass II Profile using a two-dimensional three-foot chord model, ARL Aero Note 100.

34. Küchemann, D., (1978), *The Aerodynamic Design of an Aircraft*, Pergamon Press.
35. Birkhoff, G., and Zarantonello, E.H., (1957), *Jets, Wakes, and Cavities*, Academic Press.
36. Woods, L.C., (1961), *The Theory of Subsonic Plane Flow*, Cambridge University Press.
37. Goodmmanson, L.T., and Gratzer, L.B., (1973), Recent advances in aerodynamics for transport aircraft, part 1, *Astronautics and Aeronautics*, vol. 11, no. 12, pp. 30–45.
38. Goodmmanson, L.T., and Gratzer, L. B., (1974), Recent advances in aerodynamics for transport aircraft, part 2, *Astronautics and Aeronautics*, vol. 12, no. 1, pp. 52–60.
39. Küchemann, D., (1954), Some aerodynamic properties of a new type of aerofoil with reversed flow through an internal duct, RAE TN Aero 2297.
40. Heughan, D.M., (1953), An experimental study of a symmetrical aerofoil with a near suction slot and a retractable flap, *Journal of Aeronautical Society*, vol. 57, no. 514, pp. 627–645.
41. Eminton, E., (1960), Orifice shapes for ejecting gas at the nose of a body in two-dimensional flow, RAE TN Aero 2711.
42. Edwards, J.B. (1961), Fundamental aspects of propulsion for laminar flow airfoils, in G.V. Lachmann (ed.), *Boundary Layer and Flow Control*, vol. 2, Pergamon Press, p. 1077.
43. von der Deken, J., (1971), Aerodynamics of pneumatic high lift devices, AGARD LS 43.
44. Korbacher, G.K., (1974), Aerodynamics of powered high lift systems, ARFM 6, 319.
45. Spee, B.M., (1975), V/STOL aircraft, AGARD AR-78.
46. Küchemann, D., (1947), AVA monograph, K3 MoS TR 941.
47. Küchemann, D., and Maskell, E.C., (1956), Some remarks on the jet wing and VTOL aircraft, *Aeronautical Engineering Review*, vol. 16, p. 56; RAE TM Aero 490.
48. Willie, R., and Fernholz, H., (1964), Report on the 1st European mechanics colloquium on Coanda effect, *Journal of Fluid Mechanics*, vol. 23, no.4, pp. 801–819.
49. Hagedorn, H., and Ruden, P. (1938), Wind tunnel investigation of a wing with Junkers slotted flap and the effect of blowing through the trailing edge, LGL B A 64, (also RAE LT 442, 1953).
50. Davidson, I.M., (1956), The jet flap, *Journal of Aeronautical Society*, vol. 60, p. 24.
51. Malvard, L., Poisson-Quinton, Ph., and Jousserandot, P., (1956) (in French) Recherches théorique et expérimentales sur le contrôle de circulation par soufflage appliqué aussi ailes d'avions, ONERA Note Techm 47.
52. Bauer, A.B., (1972), A new family of aerofoils based on the jet-flap principle, MDC R JS713.
53. Halsey, N.D., (1974), Methods for the design and analysis of jet-flapped airfoils, *Journal of Aircraft*, vol. 9, p. 311.
54. Shen, C.C., Lopez, M.L., and Watson, M.F., (1975), Jet-wing lifting surface theory using elementary vortex distributions, *Journal of Aircraft*, vol. 12, p. 448.
55. Küchemann, D., (1953), Some notes on the flow past aerofoils with a jet emerging from the lower surface, RAE TM Aero 399, also TM A Aero 412.
56. Spence, D. A., (1955), A treatment of the jet flap by thin airfoil theory, RAE R Aero 2568.

57. Spence, D.A. (1956), The lift on a thin jet-flapped wing, *Proceedings of the Royal Society A*, vol. 238, pp. 46–68.

58. Dimmock, N.A., (1955), An experimental introduction to the jet flap, NGTE R 175; ARC 18186, 1953.

59. Taylor, J.W.R., (ed.), (1963/1964), *Jane's all the World's Aircraft 1963–64*, McGraw-Hill Book Company, New York, p. 122.

60. Harris, K.D., (1970), The Hunting H 126 jet flap research aircraft, AGARD LS 43.

61. Butler, S.F.J., Guyett, N.B., and Moy, B.A., (1961), Six component low-speed wind tunnel tests of jet flap complete models with variation of aspect ratio, dihedral, and sweepback, including the effect of ground proximity, RAE R Aero 2652.

62. Thomas, H.H.B.M., and Ross, A.J., (1957), The calculation of the rotary lateral stability derivatives of a jet flapped wing, RAE TN Aero 2545 also ARC R & M 3277, 1962.

63. Foster, D.N., (1975), A brief flight tunnel comparison for the Hunting H 126 jet flap aircraft, RAE TM Aero 1640.

64. Photograph by Mark.murphy at English Wikipedia, https://commons.wiki media.org/w/index.php?curid=17749511

65. Küchemann, D., (1956), A method for calculating the pressure distribution over jet-flapped wing, RAE R 2573, ARC R & M 3036, 1957also TM A Aero 412

66. McCormick, B.W., (1967), *Aerodynamics of V/STOL Flight*, Academic Press, Inc.

67. Maskell, E.C., and Spence, D.A., (1959), A theory of jet flap in three dimensions, *Proceedings of the Royal Society, London, A*, vol. 251, pp. 407–425.

68. Lissamann, P.B.S., (1968), A linear theory for the jet flap in ground effect, *Journal of Aircraft*, vol. 16, no.7, pp. 1356–1362.

69. Lissamann, P.B.S., (1974), Analysis of High aspect ratio jet-flap wings of arbitrary geometry, *Journal of Aircraft*, vol. 11, no.5, pp. 259–264.

70. Lopez, M.L., and Shen, C.C., (1971), Recent developments in jet flap theory and its application in STOL aerodynamic analysis, AIAA P 71–578.

71. Kerney, K.P., (1971), A theory of the high-aspect-ratio jet flap, *Journal of Aircraft*, vol. 9, no.3, pp. 431–435.

72. Shen, C.C., Lopez, M.L., and Wasson, N.F., (1975), Jet-wing lifting-surface theory using elementary vortex distributions, *Journal of Aircraft*, vol. 12, no. 5, pp. 448–455.

73. Pankhurst, R.C., and Thwaites, B., (1950), Experiments on the flow past a porous circular cylinder fitted with a Thwaites flap, ARC R&M 2787.

74. Williams, J., and Alexander, A.J., (1955), Three-dimensional wind tunnel tests of a 30 degree jet flap model, ARC CP 304.

75. Foerthmann, E., (1933) (in German), Über turbulente Strahlausbreitung. Diss. Göttingen 1933; Ing-Arch.5, 42–54 (1934); also NANCA TM 789 (1936).

76. Sigalla, A. (1958), Measurements of skin friction in plane turbulent wall jet, *Journal of Royal Aeronautical Society*, vol. 62, pp. 873–877.

77. Barke, P., (1957), An experimental study of a wall jet, *Journal of Fluid Mechanics*, vol. 2, pp. 467–472.

78. Riley, N., (1958), Effects of compressibility on a laminar wall jet, *Journal of Fluid Mechanics*, vol. 4, pp. 615–628.

79. Launder, B.E. and Rodi, W., (1979), The turbulent wall jet, *Progress in Aerospace Science*, vol. 19, pp. 81–128.

80. Newman, B.G., (1961), The deflection of plane jets by adjacent boundaries-Coanda effect, in G.V. Lachmann (ed.), *Boundary Layer and Flow Control*, vol. 1, Pergamon Press, pp. 232–264.
81. Rodman, L.C., Wood, N.J., and Roberts, L., (1987), An experimental investigation of straight and curved annular wall jets, Joint Institute for Aeronautics and Acoustics, JIAA TR 79.
82. Young, D.W., and Zonars, D., (1950), Wind tunnel tests of the Coanda wing and nozzle, USAF Technical Report 6199.
83. Glauert, M.B. (1956), The wall jet, *Journal of Fluid Mechanics*, vol. 1, pp. 625–643.
84. Eichelbrenner, E.A., and Dumarge, P., (1962), Le problème du "jet pariétal" plan en régime turbulent pour un écoulement extérieur de vitesse Ue constante, *J. Mechanique*, vol. I, pp. 109–122 and vol. I, pp. 123–134.
85. Gartshore, J.S., and Newman, B.G., (1969), The turbulent jet in an arbitrary pressure gradient, *Aeronautical Quarterly*, vol. 20, pp. 25–56.
86. Bradshaw, P., and Gee, M.T., (1962), Turbulent wall jets with or without external stream, ARC RM 3252.
87. Kruka, V., and Eskinazi, S., (1964), The wall jet in a moving stream, *Journal of Fluid Mechanics*, vol. 20, pp. 555–579.
88. Gersten, J. (1965), Flow along highly curved surfaces, Lecture, EUROMECH I, Berlin.
89. Dovarak, F.A., (1973), Calculation of turbulent boundary layers and wall jets over curved surfaces, *AIAA Journal*, vol. 11, pp. 517–524.
90. Sforza, P.M., and Herbst, G., (1970), A study of three-dimensional incompressible turbulent wall jet, *AIAA Journal*, vol. 8, pp. 276–283.
91. Fekete, G.I., (1963), Coanda flow in a two-dimensional wall jet on the outside of a cylinder, Mechanical Engineering Department Report 63-11, McGill University, Canada (from [86]).
92. Guitton, D.E., and Newman, B.G., (1977), Self-preserving turbulent wall jets over convex surfaces, *Journal of Fluid Mechanics*, vol. 81, no. 1, pp. 155–185.
93. Gartshore, I., and Hawaleshka, O., (1964), The design of a two-dimensional blowing slot and its application to a turbulent wall jet in still air, Mechanical Engineering Department Technical Note 64-5, McGill University (from [86]).
94. McGahan, W.A., (1965), The incompressible, turbulent wall jet in an adverse pressure gradient, Reptort No. 82, Gas Turbine Laboratory, Massachusetts Institute of Technology (from [86]).
95. Bradshaw, P., (1973), Effects of streamline curvature on turbulent flow, AGARD-AG-169, August 1973.
96. Schlichting, H., Gersten, K., Krause, E., Oertel, H. Jr., and Mayes, C., (2004), *Boundary-Layer Theory*, 8th edition, Springer.
97. Launder, B. E. and Rodi, W., (1983), The turbulent wall jet – measurements and modeling, *Annual Review of Fluid Mechanics*, vol. 15, pp. 429–459.
98. Rodi, W. and Scheuerer, G., (1983), Calculation of curved shear layers with two-equation turbulence models, *Physics of Fluids*, vol. 26, no. 6, pp. 1422–1436.
99. Swamy, N.V.C., and Bandopadhay, P., (1975), Mean and turbulence characteristics of 3D wall jets, *Journal of Fluid Mechanics*, vol. 71, pp. 541–562.
100. Newman, B.G., Patel, R.P., Savage, S.B., and Tijo, H.K., (1972), Three-dimensional Jet originating from a circular orifice, *Aeronautical Quarterly*, vol. 23, pp. 188–200.
101. Prandtl, L., (1927), The generation of vortices in fluid of small viscosity, *Journal of the Royal Aeronautical Society*, vol. 31, p. 735.

102. Poisson-Quinton, Ph., (1948), Recherches théoriques et expérimentales sur le contrôle de couche limites, in VII International Congress of Applied Mechanics.
103. Wygnanski, I., (1966), The effect of jet entrainment on loss of thrust for a two-dimensional symmetrical jet flap aerofoil, *The Aeronautical Quarterly*, vol. 17, no. 1, pp. 31–52.
104. Loth, J.L., (2006), Advantages of combining boundary layer suction with circulation control high lift generation, Chapter 1, in R.D. Joslin, and G.S. Jones (eds), *Applications of Circulation Control Technologies*, Progress in Astronautics and Aeronautics, AIAA, Virginia, USA, vol. 214, pp. 1–21.
105. Cheeseman, I.C., and Reed, A. R., (1967), The application of circulation control by blowing to Helicopter rotors', *Journal of Royal Aeronautical Society*, vol. 71, no. 679, pp. 451–467.
106. Cheeseman, I. C., (1967), Circulation control and its application to stopped rotor aircraft, AIAA Paper no. 67-747.
107. Lockwood, V.E., (1960), Lift generation on a circular cylinder by tangential blowing from surface slots, NASA Langley Center, Technical Note D-244.
108. Englar, R.J., (2006), Overview of circulation control pneumatic aerodynamics: blown force and moment augmentation and modifications as applied to primarily to fixed wing aircraft, Chapter 2, in R.D. Joslin, and G.S. Jones (eds), *Applications of Circulation Control Technologies*, Progress in Astronautics and Aeronautics, American Institute of Aeronautics and Astronautics, Inc., Virginia, USA, vol. 214, pp. 23–68.
109. Englar, R.J., and Williams, R.M., (1973), Test techniques for high lift tow-dimensional airfoils with boundary layer circulation control for application to rotary wing aircraft, *Canadian Aeronautics and Space Journal*, vol. 19, no.3, pp. 93–108.
110. Englar, R.J., (1972), Two-dimensional subsonic wind tunnel tests on a cambered 30 percent thick circulation control airfoil, NSRDC, Technical Note AL-201, AD 913–411L.
111. Englar, R.J., (1975), Experimental investigation of the high velocity Coanda wall jet applied to bluff trailing edge circulation control airfoils, DTNSRDC Report 4708, Aero Report 1213, AD-A-019-417, September.
112. Englar, R.J., (1981), Low speed aerodynamics characteristics of a small fitted trailing edge circulation control wing configuration fitted to a supercritical airfoil, DTNSRDC, Report ASED-81/09, issued in March.
113. Englar, R.J., and Huson, G.G., (1984), development of advanced circulation control using high lift airfoils, *Journal of Aircraft*, vol. 21, no. 7, pp. 476–483.
114. Englar, R.J., Smith, M.J., Kelley, S.M., and Rover III, R.C., (1994), Development of circulation control technology for application to advanced subsonic transport aircraft, *Journal of Aircraft*, vol. 31, no. 5, pp. 1160–1177.
115. BBC Two (2003), Horizon – The Crash of Flight 587, broadcast on Thursday, 8 May.
116. Rossow, V.J., Fong, R.K., Wright, M.S., and Bisbee, L.S., (1996), Vortex wakes of two transports measured in 80 by 120 foot wind tunnel, *Journal of Aircraft*, vol. 33, no. 2, pp. 399–406.
117. Simpson, R.G., Ahmed, N.A., and Archer, R.D, (1999), Lift enhancement of a wing using Coanda tip jets, in *Proceedings of the 8th International Aerospace Congress*, Adelaide, September 1999. IAC99 Proceedings, Paper no. 4.3.1.
118. Simpson, R.G., Ahmed, N.A., and Archer, R.D, (2000), Improvement of a wing performance using Coanda tip jets, *AIAA Journal of Aircraft*, vol. 37, no. 1, pp. 183–184.

119. Simpson, R.G., Ahmed, N.A., and Archer, R.D, (2002), Near field study of vortex attenuation using wing tip blowing, *The Aeronautical Journal of the Royal Aeronautical Society*, vol. 102, pp. 117–120.
120. Goldscmied, F.R., (1967), Integrated hull design, boundary layer control and propulsion of submerged bodies, *Journal of Hydrodynamics*, vol. 1, pp. 2–11.
121. Chin, Y.T., Aiken, T.N., and Gates, G.S., Jr., (1975), Evaluation of a new jet flap propulsive lift system, *Journal of Aircraft*, vol. 12, no. 7, pp. 605–610.
122. Jones, G.S., (2006), Pneumatic performance of circulation control airfoil, Chapter 7, in R.D. Joslin, and G.S. Jones (eds), *Applications of Circulation Control Technologies*, Progress in Astronautics and Aeronautics, vol. 214, pp. 191–243.
123. Cerchie, D., Halfon, E., Hammerich, A., Han, G., Taubert, L-T, Varghese, P., and Wygnasnski, I., (2006), Some circulation and separation control experiments, Chapter 5, in R.D. Joslin, and G.S. Jones, (eds), *Applications of Circulation Control Technologies*, Progress in Astronautics and Aeronautics, vol. 214, pp. 113–165.
124. Carlisle, R. P., (1998), *Where the Fleet Begins: A History of the David Taylor Research Center, 1898–1998*, Department of the Navy, pp. 373–379.
125. Williams, R.N., Leitner, R.T., and Rogers, E.O., (1976), X-wing: a new concept in rotary VTOL, in presented at AHA Symposium on Rotor Technology.
126. Rogers, E.O., Schwartz, A.W., Abramson, J., (1985), Applied aerodynamics circulation control airfoils and rotors, in presented at 41st Annual AHS Forum, May.
127. https://commons.wikimedia.org/wiki/File:Sikorsky_X-wing_diagonal_view.jpg
128. Darpa ditches X-Wing, (1988), *Flight International*, 16 January issue, p. 2.
129. Photograph by Juergen Lehle, public domain: https://commons.wikimedia.org/w/index.php?curid=361153
130. Sampatakos, E.P., Morger, K.M., and Logan, A.H., (1983), NOTAR: the viable alternative to a tail rotor, AIAA Aircraft Design, Systems and Technology Meeting, American Institute of Aeronautics and Astronautics, Fort Worth, TX, AIAA Paper 83-2527.
131. Logan, A.H., Morger, K.M., and Sampatakos, E.P., (1983), Design, development and testing of the no tail rotor (NOTAR) demonstrator, 9th European Rotorcraft Forum, Stresa, Italy, paper no. 67.
132. Dempsey, M.E., (2010), Eyes of the Army—U.S. Army Roadmap for Unmanned Aircraft Systems 2010–2035, https://www.rucker.army.mil/usaace/uas/US%20Army%20UAS%20RoadMap%202010%202035.pdf
133. https://patents.google.com/patent/GB2387158A/en
134. Ahmed, R.I., Talib, A.R.A., Rafie, A.S.M., and Djojodihardjo, H., (2017), Aerodynamics and flight mechanics of MAV based on Coandă effect, *Aerospace Science and Technology*, vol. 62, pp. 136–147.
135. Lee, J.Y., Song, S.H., Shon, H.W., Choi, H.R., and Yim, W., (2017), Modeling and control of a saucer type Coandä effect UAV, IEEE International Conference on Robotics and Automation (ICRA), Singapore, May 29–June 3, pp. 2717–2722.

4

Miscellaneous Applications of Coanda Effect

4.1 Industrial and Environmental Applications

The list of areas where non-aeronautical applications of the Coanda effect is possible is very large. Here we will only look at some selected industrial and environmental cases that possess significant potential for Coanda effect applications.

Section A: Industrial Applications

4.2 Metallurgical Processes

Various metallurgical processes require greater mixing and agitation of constituent materials. In steel-making industries, for example, the commonly employed technique is gas stirring in the processing of liquids. In these instances, inert gas is introduced into a bath of molten metal. The use of inert gas is meant to ensure that it does not take part in the chemical reactions, but its upward movement facilitates larger agitation of the molten metal. The agitation helps in the homogenization of the chemical and thermal properties of the molten mixture and provides accelerated absorption of hazardous, non-metallic particles into an overlaying slag.

The gas stirring gives rise to the formation of bubbles which, depending on the arrangement or process set up, may exhibit the Coanda effect. Our goal here is to consider the way the Coanda effect takes place in bubble plumes and whether the effect could be exploited toward achieving minimum mixing time and optimum flow rates. From a manufacturing perspective, this would maximize alloy addition recoveries in a cost-effective manner in metallurgical industries.

Bubble plumes have wide ranging applications: from wave damage alleviation for building structures to prevention of channels and harbors becoming frozen, from stratification control of reservoirs and lakes for improved water quality to pollution control in oil and sea exploration. We will, however, focus on bubbles that have been found suitable in enhancing the mass and heat transfer primarily in metallurgical processes [1–6]. The difficulty arises when some of the bubbles attach to the side walls of a reactor or get trapped in the molten metals during the process with undesirable consequences. This may result in the reduction of the intensity of the mixing in the reactors or serious erosion of the reactor walls. In a metallurgical context, this may result in significant degradation [5] in the quality of the final products.

Removing the bubbles from the walls or trapped bubbles from the metals is not easy. This is because the buoyancy forces acting on them is low. Thus, there is a simultaneous requirement of generating small bubbles for agitation purposes, while at the same time ensuring that they do not attach to the walls or remain trapped. Once trapped, removing the bubbles poses significant practical challenges.

Coanda effect is likely to develop when an otherwise symmetrically developing bubble plume encounters a wall or another plume in close proximity. Under this condition, the plume stops developing further in a symmetric manner. It may bend sideways to the nearby wall or toward the second plume as the case may be.

If the plumes are laminar in nature, the pressure difference that exists between the plume and the wall or between the two plumes would be the main driver of the bending phenomenon [7]. If, on the other hand, the plumes are turbulent, then the pressure difference may play a relatively minor role, and the asymmetric plume geometry is caused by turbulent entrainment. Whatever the mechanism, the deviation of laminar and turbulent plumes from their original flow pattern is taken by most authors as the demonstration of the Coanda effect [8].

In this chapter, we will look at the flow characteristics of bubble plumes under the Coanda effect and whether the effect can be exploited for agitation while also removing them before they form an unwanted part of the structure of the metals produced.

4.2.1 Bubble Plume

Freire et al. [8] performed laboratory experiments with a plume close to a wall and also two plumes placed side by side to create the Coanda effect. They wanted to test the hypothesis that the Coanda effect or the bending of a turbulent round bubble plume is the consequence of momentum flux imbalance created by entrainment restriction imposed by the presence of a wall or another plume.

The main components of the experimental apparatus used were a glass water tank filled with a sodium chloride solution, an air injection system consisting of a mass flow meter and two injection nozzles, a two-dimensional traversing mechanism, and a data acquisition and analysis system.

In two-phase flow measurements, the difference between the electrical conductivity or resistivity of the phases is used as the basis of practical measurement by resistivity sensors. More details of the technique can be found in the References [9].

Apart from plume deflection under various gas flow rates and distances from the wall or between the two plumes, the authors also looked at the relative importance of the flow inertia to external field and surface tension in terms of modified non-dimensional Froude (F_r) and Weber (W_e) numbers, respectively defined as:

$$F_R = \frac{q^2}{gs^5}$$

$$W_e = \frac{\Delta \rho q^2}{\sigma s^2}$$

where,

- g acceleration due to gravity
- q gas flow rate of plume source
- σ surface tension
- s half distance between the plume and wall or between the two plumes
- $\Delta \rho$ density difference

For ease of understanding, Figure 4.1 is provided to show the coordinate system adopted by the authors. Here we look at some of the results from the study of Freire et al. [8] that are given in Figures 4.2–4.5.

4.2.1.1 Near a Wall

The deflection of the plume for several gas flow rates and distances from the wall are shown in Figure 4.2 in terms of x and y. This figure shows that most

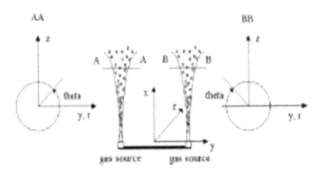

FIGURE 4.1
Coordinate system. (After: Freire, Miranda, Luz, and Franca, 2002 [8].)

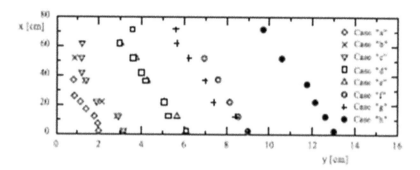

FIGURE 4.2
Deflection of plume near a wall under varying gas flow rates and distances from the wall. Here $y=0$ denotes the wall location. (After: Freire, Miranda, Luz, and Franca, 2002 [8].)

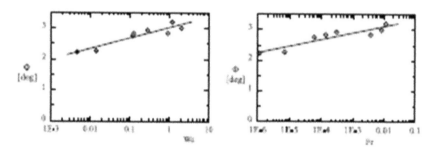

FIGURE 4.3
Deflection angle of a plume near a wall plotted against Weber and Froude numbers. (After: Freire, Miranda, Luz, and Franca, 2002 [8].)

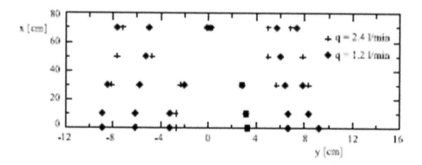

FIGURE 4.4
Deflection of two side-by-side plumes under varying gas flow rates and distances from the wall. Here $y=0$, denotes the midpoint of the two plumes. (After: Freire, Miranda, Luz, and Franca, 2002 [8].)

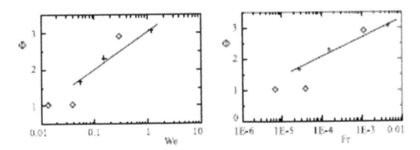

FIGURE 4.5
Deflection angle of two side-by-side plumes plotted against Weber and Froude numbers. (After: Freire, Miranda, Luz, and Franca, 2002 [8].)

of the trajectories can be approximated by a straight line, particularly when the plume distance from the wall is large. The deflection angle, shown in Figure 4.3, also shows a linear increase against the Froude and Weber numbers when plotted on a logarithmic scale.

4.2.1.2 Two Plumes Side by Side

The deflection of the two plumes is shown next in Figure 4.4, for varying gas injection rates and separation distances between the nozzles generating the plumes. The trends observed are also similar to the results shown in Figure 4.2, with the trajectories of the plume at the centerline being approximated by straight lines. The deflection angles are plotted in Figure 4.5 on a logarithmic scale against the Froude and Weber numbers, and are found to increase logarithmically with F_r and W_e, similar to those observed in Figure 4.3.

The work demonstrates the occurrence of the Coanda effect in both the cases considered and finds a clear correlation to exist between the bending angle of a bubble plume and the values of F_r and W_e that would be useful in finding the entrainment coefficient. The experimental data were also in good agreement with the results of the steady flow integral theory advanced by several authors [10, 11] providing analytical means of generating and fixing the location of the Coanda effect with appropriate flow entrainment.

Further evidence of the Coanda effect is provided more vividly in the comparative photographs shown in Figures 4.6 (a) and (b). The pictures show flow patterns for the axisymmetric case where the Coanda effect is absent and side bending plumes where the Coanda effect is present. These pictures were obtained with long exposure times to capture the mean positions of the plumes.

4.2.1.3 Bubble Characteristics

To find the bubble characteristics of bubbling jets subjected to the Coanda effect, Iguchi and Sasaki [12] conducted experiments on two-phase bubbling

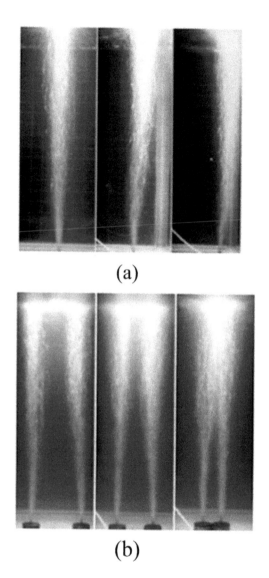

(a)

(b)

FIGURE 4.6
(a)–(b): General flow patterns showing the Coanda effect: (a) one plume close to a wall; (b) two plumes bending toward each other. (After: Freire, Miranda, Luz, and Franca, 2002 [8].)

jets rising along the sidewall of a cylindrical vessel. A schematic of the experimental set-up and the coordinate system used are shown in Figure 4.7.

Using an air–water mixture, the parameters they studied were the bubble frequency, gas hold up, mean bubble rising velocity, and mean bubble chord length. They used a two-needle, electro-resistivity probe to measure these parameters with particular attention paid to the vertical and

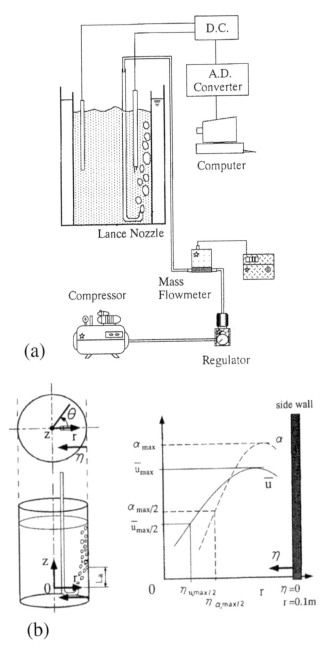

FIGURE 4.7
(a)–(b): Bubble plume subjected to the Coanda effect experiment (a). Experimental apparatus and (b) coordinate system. (After: [12].)

horizontal distributions of these quantities near the side wall. The vertical region extended from the nozzle top to the bath surface.

The study found that the attachment length, L_B, or the vertical distance from the nozzle top to the attachment position, was independent of the inner diameter of the nozzle, d_i. The attachment length, L_B, however, was dependent on both the air flow rate, Q_g, and the horizontal distance between the nozzle top and the side wall. In other words, the attachment length was independent of the inertia force of injected gas. The buoyancy force, however, exerted some influence on the length. These findings are similar to those of the Coanda effect studies in a single-jet flow [12].

Further evidence about the Coanda effect were evident from the peaks of the horizontal f_B and gas hold up distribution that shifted toward the side wall with increasing z, as shown in Figure 4.8 (a) and (b). At the same time, the mean velocity of bubble rise and mean chord length remained uniform with little change in the vertical direction as given in Figure 4.9 (a) and (b). These results suggest that the bubbles do not disintegrate or coalesce due to

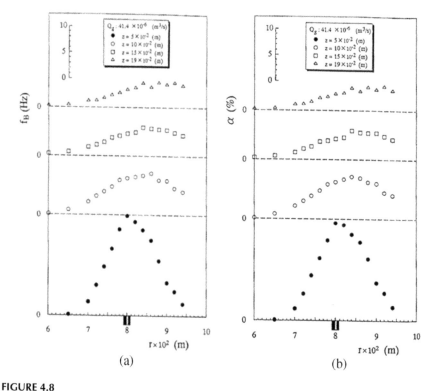

(a)   (b)

FIGURE 4.8

(a)–(b): Bubble characteristics near the side wall under the Coanda effect. (a) Bubble frequency; (b) gas hold up. (After: [12].)

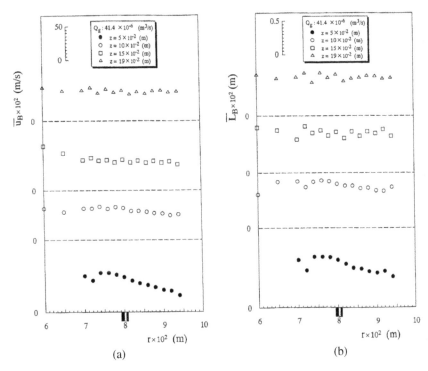

FIGURE 4.9
(a)–(b): Bubble characteristics near the side wall under the Coanda effect. (a) Mean bubble rising velocity; (b) Mean bubble chord length. (After: [12].)

the Coanda effect, and their structures remain intact as they rise along the wall.

Information regarding the mean velocity and turbulence components of water flow in and near an air–water bubbling jet under the Coanda effect has been provided by Iguchi and Sasaki [13]. They used the same experimental setup as shown in Figure 4.7 but used two-channel laser velocimetry to obtain measurements at several z-locations with particular attention paid to the horizontal distributions of the flow behavior near the vertical region above the attachment location.

The turbulence quantities at different z locations are shown in Figure 4.10 (a) and (b). The mean resultant velocity vectors near the wall obtained for a particular flow rate were plotted in the z–r plane and are shown in Figure 4.11.

In Figures 4.10 (a), (b), and 4.11, the peaks of the mean resultant velocity vector and the turbulence quantities can be seen to shift toward the cylindrical wall. From the results of this investigation, the following conclusions were drawn:

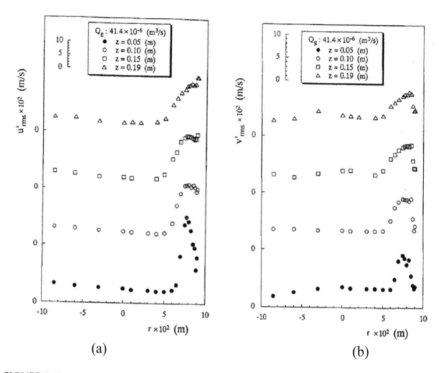

FIGURE 4.10
(a)–(b): Distribution of the r.m.s turbulence quantities (a) axial; (b) radial turbulence. (After: Iguchi, Sasaki, Nakajima, and Kawabata, 1998 [13].)

- The air–water bubbling jet, generated near the side wall of a cylindrical vessel, rose along the wall due to the Coanda effect.
- The horizontal distributions of the axial mean velocity above the attachment point were similar and not affected by the flow rate of the gas.
- The root-mean square (rms) values of the axial turbulence were of similar distribution, and a similar trend was observed for the radial turbulence distributions.
- The maximum values of axial turbulence and radial turbulence were not influenced by the sidewall and almost in agreement with their respective measured values in a vertical bubbling jet

4.2.1.4 Removal of Bubbles

We now turn our attention to the task of removing bubbles from the metallurgical processes. Several methods have been proposed to remove small bubbles from gas and liquid mixtures, such as the proposal to use pipes of

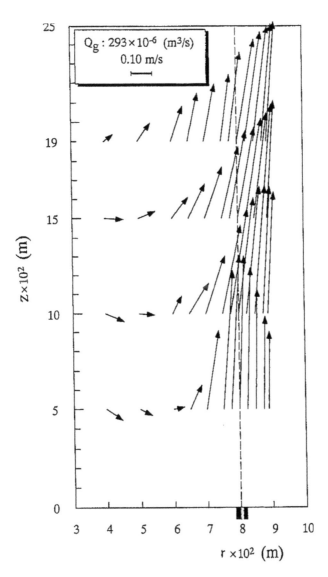

FIGURE 4.11
Resultant velocity vector distribution at a particular flow rate on a z–r plane. (After: Iguchi, Sasaki, Nakajima, and Kawabata, 1998 [13].)

poor wettability in the bath [14–16] based on the fact that bubbles in a two-phase flow are likely to attach to the inner walls of such bodies of poor wettability. Studies by Mizuno et al. [17] and Yamashita et al. [18] found that there was a critical diameter for bubbles above which such an approach would not be viable. Later works by Ogawa and Tokumitsu [19] and Iguchi et al.

[20] found that the occurrence of the Coanda effect made the requirement of the poor wettability condition redundant. They observed that if a circular cylinder was placed horizontally in a vertical water–air bubbling jet, then the spreading of the bubbles downstream of the cylinder was suppressed and bubbles attached behind the cylinder irrespective of the cylinder's wettability attribute. This fact also gave support to the conjecture that should some of the bubbles not trapped on a cylinder of poor wettability could be trapped by placing a second cylinder of poor wettability behind the first cylinder [8, 19, 20].

For further investigation, Sasaki et al. [21] investigated the flow characteristics of a two-phase flow of water–air bubbling jet downstream a circular cylinder of good wettability. To make the task of experimenting and interpreting the results better, they placed the cylinder horizontally with respect to the vertical direction of the bubbling jet.

The experimental setup [21] used is shown in Figure 4.12. Measurements were conducted with reference to a Cartesian co-ordinate system whose origin was placed on the cylinder as shown in the box of Figure 4.12.

The bubble flow behavior was investigated in terms of bubble frequency, f_B, gas hold up, α, the mean bubble rising velocity $\overline{u}_B$ and the mean bubble

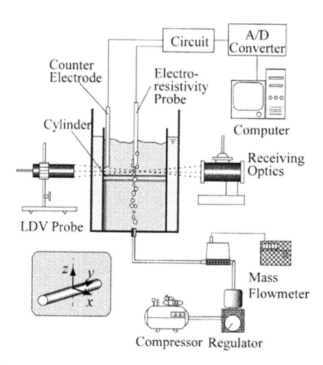

FIGURE 4.12
Experimental setup and coordinate system for bubble removing. (After: Sasaki, Iguchi, Ishi, and Yokoya, 2002 [21].)

diameter, $\overline{d_B}$. Like Iguchi and Sasaki [12], a two-needle, electro-resistivity probe was obtained to obtain these parameters. Additional information regarding the vertical, u (or axial), and v (or horizontal) velocity components were obtained using a two-channel laser Doppler velocimeter over a sample period from which their mean values, $\bar{u}$ and $\bar{v}$ were determined. The measurement methods adopted and similar details can be found in references [21–23].

The gas hold up distribution in the x- and y-directions are given in Figures 4.13 and 4.14. It can be seen in Figure 4.12 that the distribution of α appears symmetrical in relation to the z-axis, suggesting that the gas is divided equally as it flows over the cylinder. There are also two peaks

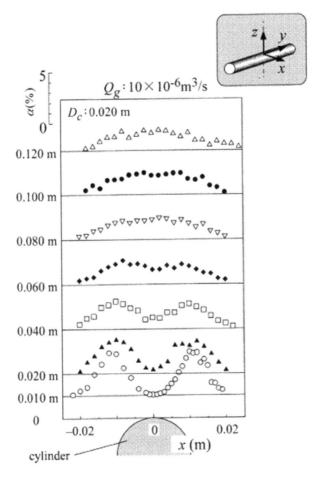

FIGURE 4.13
Gas hold up distribution behind the cylinder in the x-direction. (After: Sasaki, Iguchi, Ishi, and Yokoya, 2002 [21].)

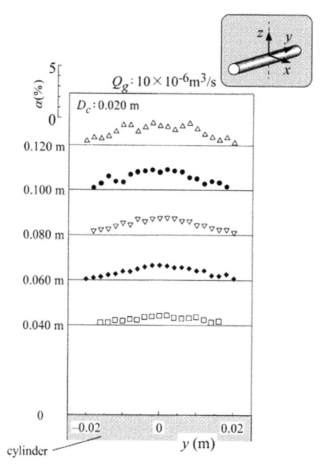

FIGURE 4.14
Gas hold up distribution behind the cylinder in the y-direction. (After: [21].)

evident behind the cylinder that gradually fade away with increases in the axial distance, z, from the cylinder.

The gas hold up distribution in the y-direction can be seen in Figure 4.14. The trend appears similar to that observed in Figure 4.13, except that the peaks are much weaker and less clear.

The mean axial and mean radial velocity distributions in the x-direction are shown in Figures 4.15 (a) and (b), respectively. Symmetric distributions and two peaks were observed in these Figures. A more useful indication of flow behavior can be obtained by examining the reduced spread of the bubbles downstream of the cylinder in the vertical direction in terms of

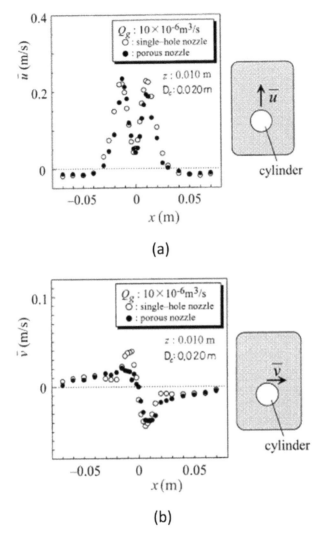

FIGURE 4.15
(a)–(b): Mean velocity distributions in x-direction (a) axial and (b) radial. (After: [21].)

the mean velocity vectors, as shown in Figure 4.16. The vector plot showed that all the velocity vectors to be directed toward the z-axis indicating flow attachment behind the cylinder. This is a demonstration of the Coanda effect not occurring when bubbles are absent in the flow [20], giving support to the notion that the Coanda effect could be used as a means of removing bubbles in metallurgical processes.

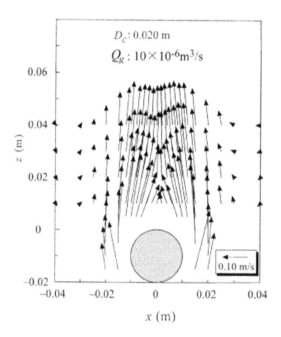

FIGURE 4.16
Mean velocity vector behind the cylinder. (After: [21].)

4.3 Grinding Processes

In the manufacturing of high-quality and high-precision mechanical and optical products, "grinding" is a vital machining process [24]. Various interactions take place during grinding, such as sliding, plastic deformation, or cutting. Malkin and Guo [25] in their thermal analysis have found high temperature grinding zones to form during the machining process of grinding. Such heat generation degrades the quality of surface finish and requires substantial coolants to be used. However, the rotating boundary of the wheel surface hinders the introduction of adequate amount of coolant fluid into the required contact zone [26]. In other words, a large amount of fluid is wasted during the coolant application.

In most manufacturing sites, air scrappers are generally used to eliminate the formation of rotating boundary layers and to provide cooling in grinding with reduced grinding fluid usage. The function of the air scrappers, however, depend to a large extent on the design of the nozzles used in supplying the grinding fluid. There are various nozzles that have been developed for this, such as coherent jet nozzle, spray nozzle, floating nozzle [27–30], and so forth. While the grinding performance using these nozzles in terms of cooling has been satisfactory, the requirement of a lower amount of coolant fluid has posed considerable challenges. This has resulted in various efforts toward developing strategies [31] for coolant use and optimization of its supply [32–34], of which the Coanda effect nozzle is one.

4.3.1 Flexible Brush-Nozzle

Hosokawa et al. [35] have reported the development of a contact-type flexible brush-nozzle that is claimed to substantially reduce the consumption of cooling fluid. The nozzle is intended to be effective in thermally unopened grinding, particularly in configurations such as a crankshaft or gear. Such a configuration is shown in Figure 4.17. The nozzle design is based on the "Coanda effect."

The basis of the design of the Coanda effect brush-nozzle is to use the Coanda effect to make the grinding fluid cling to the wheel surface during rotation thereby reducing the grinding fluid consumption. The authors were interested in examining the case of cylindrical plunge grinding and the effect of the flow around the wheel surface. They performed both computational fluid dynamic analysis and experimental investigation to validate their design and its performance.

4.3.2 Numerical Investigation

A commercial software package called ANSYS FLUENT with a k-ε turbulent model was used. The numerical model consisted of a wheel with a prescribed surface irregularity, wheel gap, and rotational speed. Figure 4.18 shows the flow field obtained around the rotating wheel due to the air scraper. The numerical study suggests that the boundary layer can be eliminated by the scrapper when the gap is zero, but reappears immediately thereby requiring the simultaneous injection of cooling fluid.

The effect due to wheel surface irregularity is shown in Figure 4.19. The non-dimensional plot of radial velocity distribution clearly shows that, with any decrease in surface irregularity, the boundary layer also decreases, but is not completely removed.

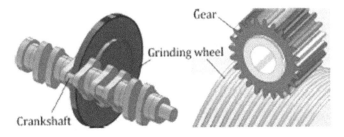

FIGURE 4.17
The nozzle design is based on the "Coanda effect."

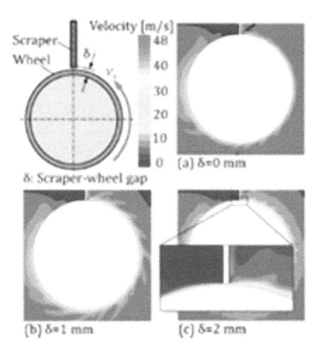

FIGURE 4.18
Effect of air scraper on air flow around the rotating wheel. (After: [35].)

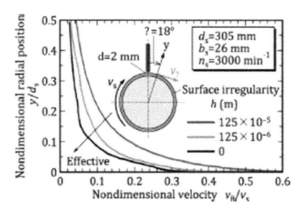

FIGURE 4.19
Effect of wheel surface irregularity on air flow around the rotating wheel. (After: [35].)

These results supported the view that a gapless scraper with a suitable amount of coolant fluid for grinding operation may work, and inspired the authors to develop two versions of internal supply and external supply contact-type fluid brush-nozzles as shown in Figures 4.20 (a) and (b).

4.3.3 Grinding Experiments

The authors conducted cylindrical plunge grinding experiments on a typical cylindrical grinding machine and observed an air scraper and both versions of the Coanda brush-nozzle. The results were encouraging. Both versions of the brush-nozzle operating on the principle of Coanda effect were found capable of making the grinding fluid adhere to the rotating grinding wheel, even at a very low grinding fluid flow rate of 0.7 liters/min. For the conventional air scraper, on the other hand, the grinding fluid was found to detach

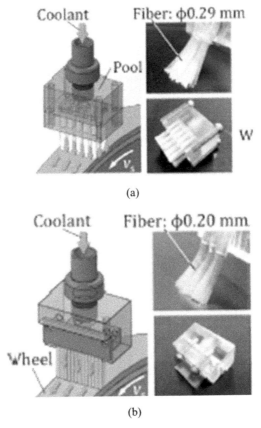

(a)

(b)

FIGURE 4.20
(a)–(b): Coanda brush-nozzle (a) internal supply; (b) external supply. (After: [35].)

from the wheel surface even when the gap was as small as 0.1 mm, becoming less effective in cooling operation. The experimental setup, the air scraper, and the Coanda effect brush-nozzles in operation are shown in Figure 4.21.

Hosokawa et al. [35] also used plunge grinding with a fiber-coupled, two-color pyrometer to evaluate the performance of the grinding characteristics of the Coanda effect brush-nozzle in terms of grinding force, grinding temperature, and surface roughness. The results suggested that the Coanda effect nozzles were effective in grinding operation and required less grinding cooling fluid. A schematic of the experimental arrangement used is shown in Figure 4.22.

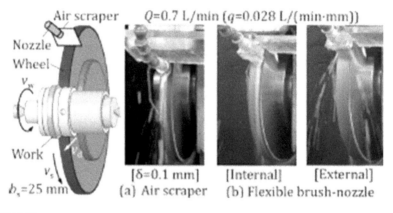

FIGURE 4.21
Experimental setup (on the left) and the air scrapper and Coanda effect brush-nozzles in operation. (After: [35].)

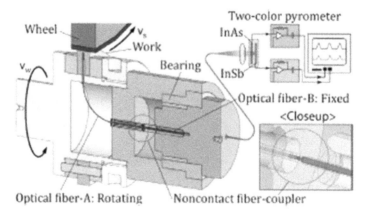

FIGURE 4.22
Experimental arrangement of a pyrometry in plunge grinding. (After: [35].)

4.4 Heat and Mass Transfer Process

Heat and mass transfer are important considerations of heat treatment in industrial processes, such as cooling of material in metal forming, cooling of electronic components, defogging of optical surfaces by heating, cooling of turbine blades, aerodynamic stabilization of floating strips, and so forth. The jet systems may be introduced normally or tangentially to the surface.

A number of reviews [36–41] have discussed various aspects of an impinging jet. The impinging jet generally emerges from a nozzle or opening. Its flow characteristics, such as the velocity and temperature profile or turbulence properties, are, however, greatly influenced by the upstream flow. Figure 4.23 shows a simplified description of some of the features of a free jet impinging normally on a surface.

If the target surface is moving and has uneven surfaces, the jet nozzles have to be located at a significant distance, which decays the kinetic energy available in the jet and reduces the Nusselt number, or cooling efficiency, of the jet significantly. In situations where the target surface is static and uniform, constructing and implementing a high-density jet at a small height may be difficult, and the efficiency of the jet may degrade because of lower jet to jet interactions.

The above considerations make a good case for establishing Coanda flow that will remain attached for a greater portion of the target surface. The process can be aided if the surface is curved and has round or cylindrical

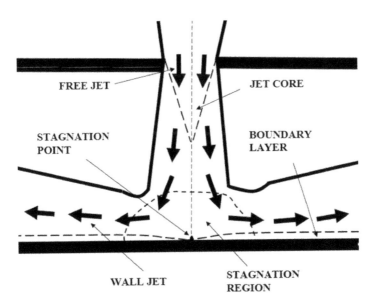

FIGURE 4.23
Schematic of a free jet impinging normally on a surface.

surfaces. Kramer et al. [42] investigated three such cases where the Coanda effect has provided enhanced performance. We will briefly describe these applications.

4.4.1 Drying of Warp Threads

Kramer et al. considered [42] the performance of a dryer to help dry warp threads by fitting the steam-heating cylinders used in drying with a nozzle system that produced jets tangentially. The drying efficiency of the dryer was found to improve substantially. Another benefit of the approach was that a large portion of the surface was reached without any ducts or vanes. Figure 4.24 shows a schematic of the application of the Coanda effect in a dryer.

4.4.2 Heating of Circular Section Logs

The conventional heating system of logs with a circular cross section is shown in Figure 4.25 (a). The logs are placed very close together with very little spaces between them to minimize the furnace length. Therefore, only a small circumference of the logs is exposed to convective heating. The situation was improved considerably by introducing nozzle slots to create the Coanda effect. The conventional heating and heating by Coanda effect are shown in Figures 4.25 (a) and (b).

4.4.3 Quenching of Metals

In this example, two-phase flow in the form of an air and water mixture is used to generate Coanda effect. Figure 4.26 shows the sketch of a Coanda

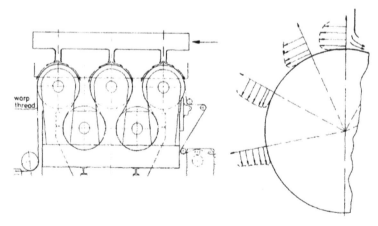

FIGURE 4.24
Coanda effect application to improve the efficiency of a dryer. (After: [42].)

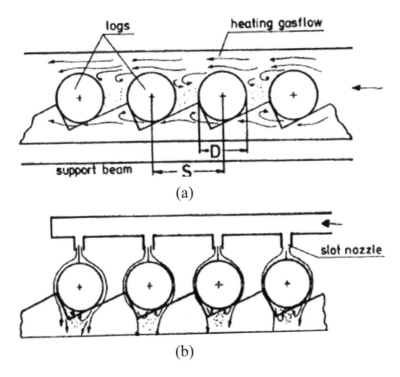

FIGURE 4.25
(a)–(b): Coanda effect in cylindrical log heating (a) conventional method (no Coanda effect);
(b) Coanda effect applied. (After: Kramer, Gerhardt, and Knoch, 1984 [42].)

device, which is made up of an air duct with an equally distributed array of
nozzles and a water tube running through its center. The water flows with
a velocity lower than the airflow. The water droplets are accelerated by the
airflow in the nozzles, which have a constriction at their ends and impinge
on the target metal sheet or metal plate. The Coanda surface placed near the
nozzle exit prevents the backflow.

Evidence of the effectiveness of the two-phase flow Coanda device
as depicted in Figure 4.26 can be obtained by observing the flow patterns shown in Figure 4.27 (a) and (b). As the water droplets move and
impinge on the metal surface, backflow occurs. This can be clearly seen in
Figure 4.27 (a). However, the presence of the Coanda surface helps move
the water droplets by the airflow without the water droplets touching
the work surface, as can be seen in Figure 4.27 (b). This induces a very
sharp start of the cooling process, which is significant in quenching. By
changing the water pressure or water droplet diameter, the device is able
to vary the cooling intensity and control the mass flow ratio of two-phase
flow constituents.

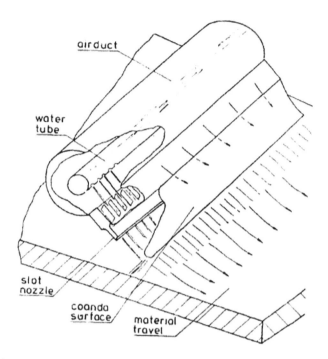

FIGURE 4.26
Sketch of a Coanda device employing two-phase flow. (After: Kramer, Gerhardt, and Knoch, 1984 [42].)

4.5 Robotic Handling

Robots are increasingly used as part of automation in modern day industries to replace many of the tasks performed by humans. Advanced gripping of objects of complex shapes and properties by robots is growing in importance in various areas, such as the textile and leather industries, micromechanical systems, surgical applications, and so forth. Figure 4.28 shows a robot hand [43] gripping or holding an object, in this case, a lightbulb.

The task of designing grippers is not easy. The surface, shape, and structure of an object are the three main requirements that dictate the suitability of universal gripping.

There are various gripping methods that can be used [44–48]. Suction gripping, needle gripping, freeze gripping, and clamp gripping are the most common. Suction gripping works well on low to non-permeable materials but is not very effective on porous bodies or materials that have uneven surfaces. Needle or freeze type grippers can leave unwanted marks on smooth surfaces. Clamp gripping, on the other hand, can pinch or distort a material.

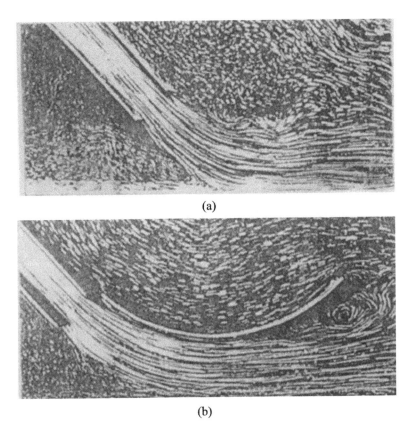

(a)

(b)

FIGURE 4.27
(a)–(b): Pictorial evidence of a Coanda device employing two-phase flow with flow flowing from left to right; (a) no Coanda surface, backflow can be observed; (b) Coanda surface added, no backflow is observed. (After: Kramer, Gerhardt, and Knoch, 1984 [42].)

In other words, the task of designing universal grippers is a difficult one. The task becomes even more difficult in the case of limp materials used in the textiles and leather industries.

The Coanda ejector as a vacuum generator appears to offer a promising solution to the gripping problem for limp materials. Various aspects of the flow field and design rules have been explored by Ameri and Dybbs [49] using the Coanda principle.

4.5.1 Working Principle of a Coanda Ejector

The working principle of a Coanda ejector is based on accelerating a pressurized jet of air through an annular nozzle before it expands via a diffuser without the flow detaching from the walls of the diffuser. This is shown in Figure 4.29.

FIGURE 4.28
A robot hand gripping an object (lightbulb). (After: [43].)

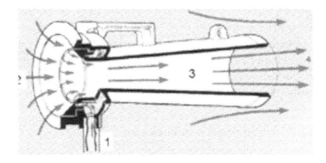

FIGURE 4.29
Working principle of Coanda ejector. (After: [50].)

4.5.2 Cylindrical and Planar Coanda Ejectors

A picture of a cylindrical Coanda ejector prototype is shown in Figure 4.30.

However, such cylindrical Coanda ejectors may not be very effective when it comes to picking up stacks of textiles or other limp materials from the shelves, where the size of the cylindrical Coanda ejectors may prove to be large. Lien and Davies [50] proposed slimmer planar Coanda ejector to overcome the size issue. These were lateral Coanda ejectors that had a rectangular configuration as shown in Figure 4.31 (a).

Comparative performance tests were conducted on the cylindrical and planar Coanda ejectors. The results showed that the planar Coanda ejectors performed better, and multiple head grippers can be developed. The results also provided some basic design guidelines about the ejector that are summarized below:

- The diffuser length should be eight times larger than the diffuser width.
- The height to width ratio of the diffuser should be two.
- The diffuser channel should be narrow for leather materials but wider for porous materials.

Lien and Davies [50] also suggested building multiple head suction grippers as shown in Figure 4.32, and using an array of single-sided planar Coanda ejectors with common air supply. Each ejector of the multiple head gripper can then be used as a suction head, independent of the others, to pick up materials with the desired lifting force.

Apart from limping materials, the Coanda ejector can also be applied in the food industry to pick up fish fillets or meat pieces. The fact that the Coanda

FIGURE 4.30
Cylindrical Coanda ejector. (After: [50].)

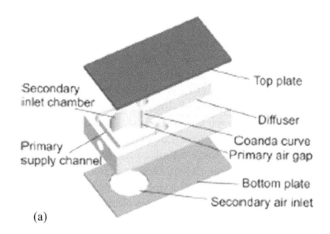

Secondary
inlet chamber

Top plate

Diffuser

Primary
supply channel

Coanda curve
Primary air gap

Bottom plate
Secondary air inlet

(a)

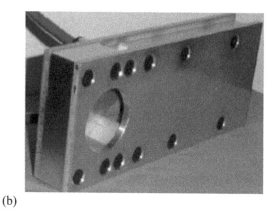

(b)

FIGURE 4.31
(a)–(b): Planar Coanda ejector (a) computer aided design description and (b) prototype.
(After: [50].)

ejector provides self-cleaning makes it a useful, hygienically safe device for
operation in food industries.

Section B: Environmental Applications

4.6 Gas Waste Flares

Methane is among the most powerful greenhouse gases arising from oil
and gas explorations. Although methane makes up of only 9% of greenhouse gases, its release to the atmosphere can be extremely damaging.

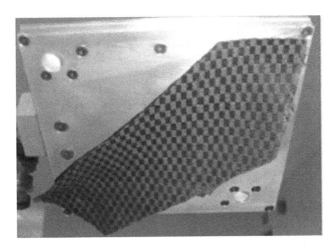

FIGURE 4.32
Two heads of a four ejector Coanda gripper picking up a textile. (After: [50].)

This is because methane's estimated global warming potential is over 100 times greater than that of carbon dioxide, CO_2 [51]. Although CO_2 is also a contributor to global warming and may stay in the atmosphere for a longer period than methane, it probably still makes sense to convert methane to CO_2 through a gas flare before releasing it into the atmosphere so as to lower environmental harm. Figures 4.33–4.36 show pictures of flaring.

FIGURE 4.33
North Dakota flaring of gas. (After: Doubek [52].)

FIGURE 4.34
Flaring of associated gas from an oil well site in Nigeria. (After: Chebyshev [53].)

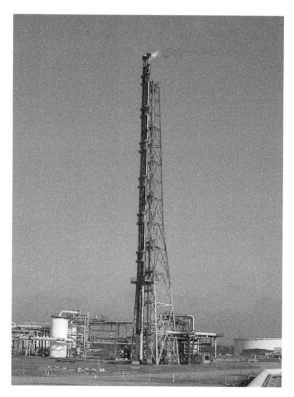

FIGURE 4.35
Flare stack at the Shell Haven refinery in England. (After: Terryjoyce [54].)

FIGURE 4.36
Flaring gases from an oil platform in the North Sea. (After: [55].)

A gas flare, also known as a flare stack, is used in the combustion of flammable gas in industrial chemical and gas processing plants, and oil refineries. The flaring is intended to protect against over-pressuring industrial plant equipment from catastrophic incidents. A flare is also used at oil or gas production sites that have oil wells, gas wells, and so forth for wastes or unusable gas.

Improper flare operation may emit methane and other volatile compounds that can cause respiratory and other health problems. Flaring is also life threatening to wildlife as it attracts birds and insects to the flame. According to CBS News [56], over 7,500 migrating songbirds were attracted to and killed by the flare at the liquefied natural gas terminal in Saint John, New Brunswick on September 13, 2013. Similar tragic incidents have also been reported for flares on other offshore oil and gas installations [57].

A report published in *Premium Times* [58] mentions $850 million being lost in Nigeria in 2015 alone due to gas flare. According to a report by World

Bank [59], until 2011, the annual amount of gas flared amounted to nearly 150×10^9 cubic meters (5.3×10^{12} cubic feet); this is equivalent to about 25% of the annual natural gas consumption in the United States or about 30% of the annual gas consumption in the European Union [59], whose total market value, if calculated at \$5.62 per 1000 cubic feet, is nearly \$30 billion [60]. These are huge amounts of losses in economic and environmental terms for both developed and developing countries and emphasize the need for technologies that produce lower volumes of waste gas, as well the adoption of air pollution abatement strategies before exhausting. These are important issues but further discussion on them is beyond the scope of this book.

We will look at waste gas flares from the Coanda effect application point of view. In the late 1960s, the British Petroleum Research Center started exploring the concept of new burners based on the Coanda effect concept [61–65] to entrain air into the gas supply of pre-mixed burners. This has led to the development of various versions of the Coanda waste gas flare, such as the Indair, Stedair, or Mardair. All these types are smoke-free, low radiation flare systems.

The development of the Coanda effect gas flare began with experiments [65] conducted on internal venturi pumps with a very small throat diameter measuring only a few inches. The device throat had a slot that faced radially inwards. A laminar jet of fuel gas at sonic velocity was ejected through the slot to produce the Coanda effect by producing a low-pressure region on either side of the jet. The low-pressure region between the jet and the venturi wall caused the jet to attach to the venturi wall. On the other side of the jet, the low-pressure region created entrainment of flow from its surroundings, forming a core of air that accelerated through the center of the venturi and exited, imparting a change in the direction of the jet.

These were very promising results since entrainment of air volume as high as 20 times per volume of fuel jet was demonstrated. This high entrainment feature also essentially renders it become a Coanda jet pump or ejector for ventilation purposes in coal mines or other hazardous environments. For combustion, the results meant that the process was capable of producing a near stoichiometric ratio that was conducive to the generation of a compact, non-luminous flame upon ignition.

Further work, therefore, pursued external Coanda effect gas flares. This involved topological inversion and took the form of a Coanda surface where the jet gas was issued from the base of the tulip through an annular exit slot that was facing radially outward.

The results obtained were similar to the venturi type of burner mentioned above. The divergence and curvature of the flow produced high entrainment of ambient air and consequently good mixing.

The external Coanda effect gas flare is considered more practical because it can be scaled up and manufactured with greater ease and, therefore, has found widespread use.

A further attraction of a Coanda device is the low level of noise associated with a Coanda flow [64, 66, 67]. Since the Coanda effect waste gas flare is considered a high-speed device, Carpenter and Green have conducted theoretical and experimental investigations on the acoustic characteristics [67] of such a device. For experimental study, they used a 1:8 scale model of a 33-inch commercial Indair waste gas flare manufactured by Kaldair Ltd. The Indair flare is generally fitted with a step between the nozzle exit and the Coanda surface to promote greater mixing and ensure flow attachment.

The authors carried out detailed studies of the flow field associated with the tulip shaped Coanda model, using spark Schlieren, shadow graph, and surface oil-film methods. Acoustic measurements were also carried out in an anechoic chamber. Two configurations were considered, one without a step and the other with a 2 mm step at the same stagnation pressure ratio of $\frac{p_0}{p_\infty}$ = 5.2. We will highlight the main features of the flow field in both cases.

4.6.1 Coanda Indair Flare without Step

Figure 4.37 shows flow visualization using surface oil-film methods. Figures 4.38 (a) and (b) show spark Schlieren photographic details of region 4 of Figure 4.37.

FIGURE 4.37
A surface oil-film flow visualization on a Coanda Indair flare model with steps. Key: 1, sleeve to vary slot width; 2, exit slot; 3, Coanda surface; 4, flow field, details in Figure 4.38; 5, separation bubble, and 6, streamwise vortices. (After: Carpenter and Green, 1997 [67].)

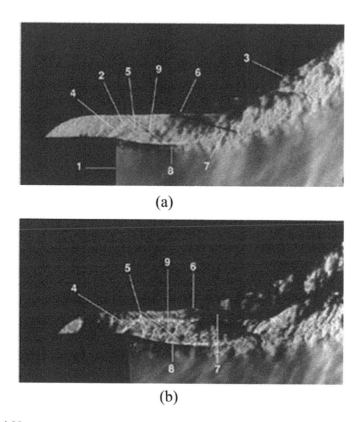

(a)

(b)

FIGURE 4.38
(a)–(b): Details of the region marked 4 of Figure 4.37 using spark Schlieren Key: 1, adjustable sleeve; 2, expansion wave fan; 3, Coanda surface; 4, internal shock wave; 5, lip shock; 6, separation bubble at leading edge; 7, separation shock wave; 8, edge of jet; 9, reflected shock wave. (After: Carpenter and Green, 1997 [67].)

The results show the development of an axisymmetric supersonic wall downstream of the exit slot. This jet that forms on a Coanda surface is made up of an outer mixing layer, a potential core, and a boundary layer. The shock wave system encountered is highly complex. The internal shock waves and separation bubbles formed on the outer nozzle walls are possibly due to the sign change of the streamline curvature. The resulting wave system is, therefore, more complex than that observed in round jets. The internal shock waves of the nozzles, however, are reflected on the jet and interact with the flow that is underexpanded in most cases, forming a somewhat quasi-periodic structure of shocks and expansions in the Coanda flow. The waves may also be formed by a radial expansion of the waves. The formation of separation bubbles, which grow in size with increasing operating pressure before finally breaking away, are the consequence of interactions of the shock waves with boundary layer on the Coanda surface.

The acoustic measurements showed that discrete tones are present in the sound pressure level (SPL) spectra, particularly at high operating pressure (6.52), and are related to the formation of large separation bubbles. In this paper, the authors also advanced the concept of a self-excited feedback loop as the mechanism for the generation of tones.

For ease of understanding, the authors also provided sketches to depict the flow features of the high speed Coanda flow. These are shown in Figures 4.39 through 4.42.

4.6.2 Coanda Indair Model with Step

Figure 4.43 shows flow visualization using surface oil-film methods. Figure 4.44 (a)–(b) show spark Schlieren photographic details of region 2 of Figure 4.43.

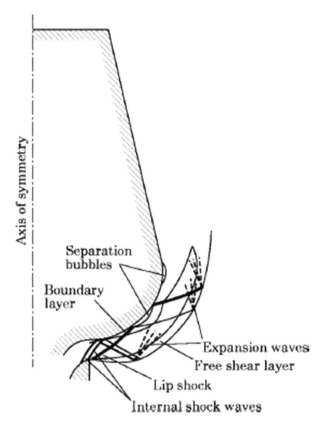

FIGURE 4.39
Key features of the Indair flow field without a step. (After: Carpenter and Green, 1997 [67].)

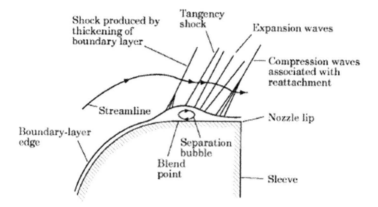

FIGURE 4.40
Mechanism of the shock wave generation at nozzle's exit. (After: Carpenter and Green, 1997 [67].)

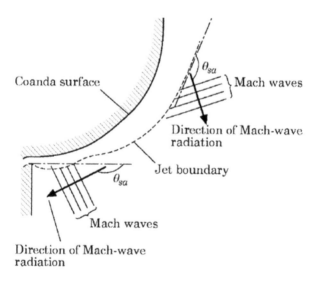

FIGURE 4.41
Mach wave radiation. (After: Carpenter and Green, 1997 [67].)

Additional tests were also conducted by replacing the annular exit slot with a saw-toothed one. The main findings of Carpenter and Green [68] can be summarized as follows:

- A region of base flow is formed immediately downstream of the step that substantially alters the flow at the exit and eliminates the lip shock that was observed in the unstepped case. This confirmed the

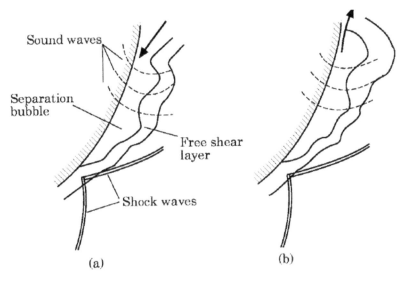

FIGURE 4.42
(a)–(b): Generation of discrete tones (a) inflow to bubble; (b) outflow from bubble. (After: Carpenter and Green, 1997 [67].)

FIGURE 4.43
A surface oil-film flow visualization on a Coanda Indair flare model with steps. Key: 1, sleeve to vary slot width; 2, exit slot; 3, Coanda surface; 4, flow field, details in Figure 4.44; 5, step between exit slot and Coanda surface. (After: Carpenter and Green, 1997 [68].)

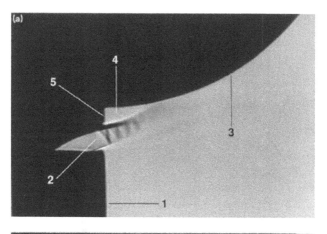

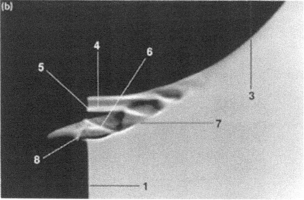

FIGURE 4.44
(a)–(b): Details of the region marked 4 of Figure 4.43 using spark Schlieren. Key: 1, adjustable sleeve; 2, annular exit slot; 3, Coanda surface; 4, base flow downstream of step; 5, step between exit slot and Coanda surface; 6, lip shock; 7, jet boundary; 8, internal shock wave. (After: Carpenter and Green, 1997 [68].)

notion that the incorporation of a step delays flow separation to a higher operating pressure. This conclusion is also supported by the work of Gregory-Smith and Senior [69].

- Two groups of discrete tones are generated by the presence of a step, one at relatively low stagnation pressure where the dominant frequency does not change with changes in stagnation pressure, while the other at a high stagnation pressure where the dominant frequency falls with increases in stagnation pressure.

- The saw-toothed configuration enhances entrainment and generates streamwise vortices that have the effect of disrupting the coherent structures by reducing their scale and shock-cell structure.

As with the case of a flow field without any step, Carpenter and Green's explanations regarding the above findings on the flow features are provided in terms of simplified sketches and shown in Figures 4.45–4.48.

4.7 Premixed Flame Stabilization

We will now consider stabilization and the flow characteristics of pre-mixed flames using the Coanda effect in an axisymmetric curved wall jet. Premixing fuel and air in lean proportions can lower peak flame temperatures and thereby reduce harmful NOx emissions.

Since a jet flow is employed mostly in the design of burner systems for practical applications, flame stabilization is an important consideration to enhance combustion load and ensure burner safety.

The enhancement of combustion load requires greater mixing of the flow, which is often achieved by creating a recirculating zone. Such recirculation of the flow helps to stabilize the flame. A recirculating region can be formed in the wake of a body. The flame holder is generally used to act as a bluff body. Other methods involve the use of swirl flow or sudden expansion of the combustor wall. All these methods, however, have their limitations and

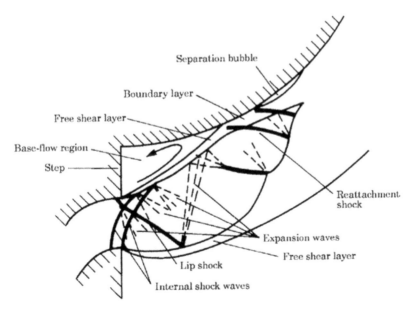

FIGURE 4.45
Key features of the Indair flow field with a step. (After: Carpenter and Green, 1997 [68].)

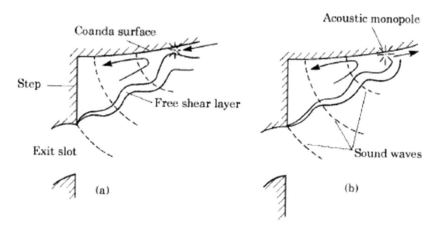

FIGURE 4.46
Mechanism of generating discrete tones at low stagnation pressures. (After: Carpenter and Green, 1997 [68].)

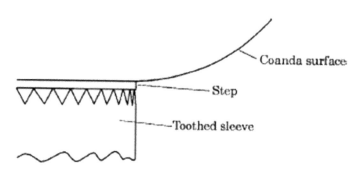

FIGURE 4.47
Sketch of a triangular shaped. saw-toothed exit slot. (After: Carpenter and Green, 1997 [68].)

can create undesirable features such as an excessive loss of momentum in the axial direction.

For burner safety, it is important to prevent flashback in premixed burners. Quenching of the flame and the mixing of the fuel with air close to the nozzle exit can help prevent flashback. The excessive heating of the combustor wall can be prevented by controlling the flame length and reducing the heat transfer between the flame and heat exchangers.

Gaydon et al. [70] note that the divergence and the curvature of a flow produce rapid entrainment of the ambient air in high premixing. We have also seen confirmation of this effect in the works mentioned in Section 4.5.

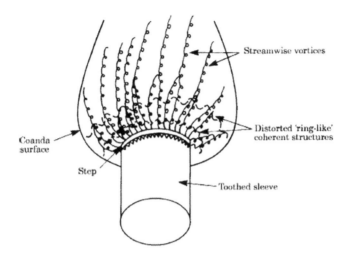

FIGURE 4.48
The effect of streamwise vortices on the coherent structures near the exit slot region. (After: Carpenter and Green, 1997 [68].)

4.7.1 Coanda Burner

Gil et al. [71] investigated an axisymmetric curved wall jet configurations such that an axisymmetric, radially inward jet was generated. The working principle of the burner they proposed can be explained using Figure 4.49.

The streamline curvature creates low pressure regions on both sides of the jet resulting in flow clinging to the surface of the cylinder and while entrainment of air takes place on the ambient side of the jet producing the Coanda effect. Downstream of the cylinder, the pressure eventually recovers and becomes larger than the ambient pressure near the center of the cylinder, at which point the flow separates from the cylinder forming a recirculating zone. Outside this recirculating zone, the jets from the two sides of the cylinder meet forming the jet interaction region. The jets finally merge and exhibit the properties of self-similar flows.

The authors investigated the flame height, blow-off, and extinction characteristics to form the basis of the design of an axisymmetric curved-wall jet burner using an experimental apparatus that consisted of a an axisymmetric curved-wall jet burner and a flow control system. Information was gathered from flow visualization pictures using direct photography, A Schlieren system and laser Doppler velocimeter operated in back scatter mode.

The arrangement of the axisymmetric curved-wall jet burner is shown in Figure 4.50. It is made up of an inner cylinder with a half sphere, an outer cylinder with an outer nozzle guide, and an inner cylinder adjustor.

Flames' regimes can be broadly examined under three conditions. Figures 4.51 (a)–(c) show direct photographs of three cases:

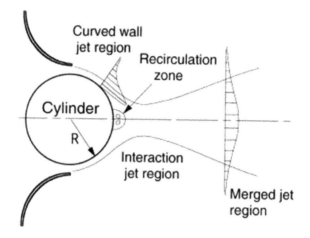

FIGURE 4.49
Working principle of an axisymmetric curved wall jet Coanda burner. (After: [71].)

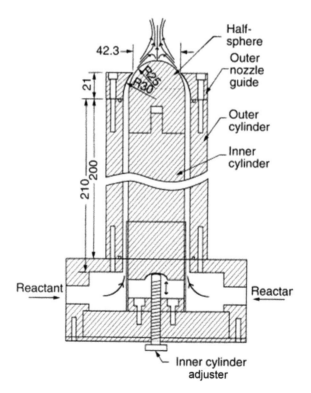

FIGURE 4.50
Arrangement of axisymmetric curved-wall jet burner. (After: Gil, Jung, and Chung, 1998 [71].)

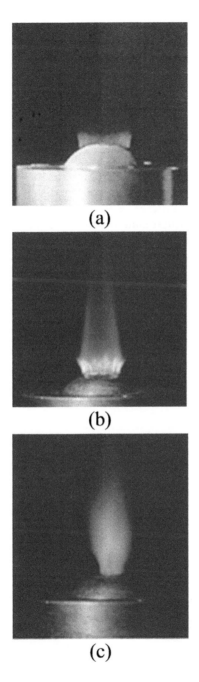

FIGURE 4.51
(a)–(c): Different flame regimes: (a) stoichiometric premixed laminar; (b) rich premixed laminar; (c) premixed turbulent. (After: Gil, Jung, and Chung, 1998 [71].)

- Stoichiometric premixed flame in the laminar regime: the flame has a smooth surface with a crown shape. The authors found the balance of the flow velocity and flow-burning velocity to stabilize the flame.

- Rich premixed flame in the laminar regime: the lower part of the flame is also of a crown shape, but the surface is corrugated. The authors attributed this to the cellular instability arising from the differences between the rates of thermal and mass diffusion. The effect of the Görtler instability from streamline curvature was found to be minimal.

- Premixed flame in the turbulent regime: as the flow rate is increased, the stabilization of the flow happens at a distance further away from the exit. With further increases in flow rate, the flow eventually transitions from the laminar state turbulent regime. There is a decrease in flame height and any distinction between the premixed flame and the diffusion flame becomes difficult to distinguish.

4.7.2 Experimental Results

Figures 4.52 (a) and (b) show the Schlieren pictures of rich propane jets for laminar and turbulent regimes, respectively. In Figure 4.52 (a) the cellular structure is evident in the laminar regime. Small scale turbulence is not evident in the jet interaction region, but present in the merged region. In Figure 4.52 (b), the jet spread angle appears much larger, about three times larger than the premixed flames with an annular spread angle observed by

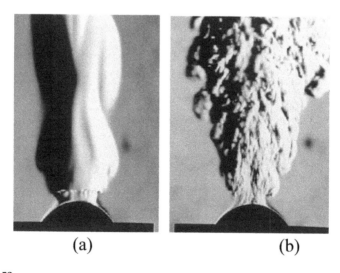

(a)   (b)

FIGURE 4.52
(a)–(b): Schlieren pictures of premixed flames: (a) laminar regime; (b) turbulent regime. (After: Gil, Jung, and Chung, 1998 [71].)

Lewis and von Elbe [72]. Higher entrainment and the effect of accelerated velocity from the upper region of the recirculating zone were thought to have caused the larger spread angle.

Figure 4.53 shows the velocity field and streamlines drawn from laser Doppler measurements near the burner, including the recirculating zone. The results show that the flame is kept inside the recirculating zone. The flow velocity is small in this region and the flame velocity can be maintained at a higher velocity than the blow-off velocity of the tube.

In conclusion, it may be said that the work of Gil et al. [71] demonstrated the advantages of the axisymmetric curved wall jet burners in terms of extending the blow-off range, reducing the flame height, and providing greater control of the nozzle exit area. This suggests that the Coanda effect burners were more effective in achieving the practically significant task of flame stabilization than conventional tube jet burners.

4.8 Coastal Engineering

Another important area where the Coanda effect occurs is during the interactions between coastal water flows and marine structures. The stream–structure interaction is a hydrodynamic feature that has high environmental impacts when it comes to the design, operation, and maintenance of the quality of shallow waters and structures of a harbor. In Figures 4.54 and 4.55,

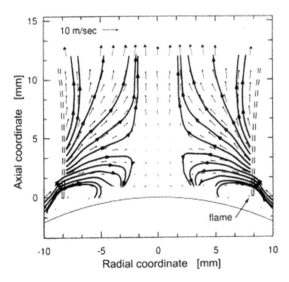

FIGURE 4.53
Laser Doppler measurements near the burner and the recirculating zone. (After: Gil, Jung, and Chung, 1998 [71].)

FIGURE 4.54
Aerial photo of Sydney Harbor. (After: Haywood [73].)

FIGURE 4.55
Aerial view of Latina Harbor. (After: Public domain [74].)

aerial views of Sydney Harbor, Australia [73] and Latina Harbor, Italy [74] are shown.

In shallow coastal waters, the accompanying diffusion processes that take place between the water flows from the sea or rivers heading toward the sea and stationary or moving marine structures lead to changes in flow patterns along the coast. Other hazardous scenarios are created by the presence of features of sandbars by peers, jetties and so forth, particularly when wind and breaking waves push surface water toward the land and form strong currents, such as rip currents that are strong, localized, and narrow. Rip currents move directly away from the shore in a manner similar to a river flowing out to sea. Figure 4.56 shows a diagrammatic representation of a rip current and a warning sign on a beach and what to do when caught in a rip current.

From a fluid mechanical point of view, the fluid–solid interactions and their effects on the coastal banks may be likened to the behavior of jets impacting on a wall or jets being subjected to cross flow. There have been various numerical and experimental studies conducted of jet wall interactions in shallow waters [76–85].

In numerical simulations, it is worthwhile to note that the two-dimensional conservative equations used for shallow water are generally depth averaged and take the following form [85]:

$$\frac{\partial \eta}{\partial t} + \frac{\partial U}{\partial x} + \frac{\partial V}{\partial y} = 0$$

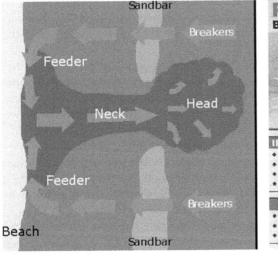
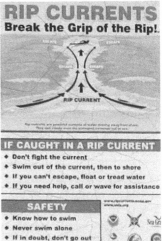

FIGURE 4.56
Schematic of rip current and warning sign posted at Mission Beach, San Diego. (After: Commons [75].)

$$\frac{\partial U}{\partial t} + \frac{\partial \left(\frac{U^2}{H}\right)}{\partial x} + \frac{\partial \left(\frac{UV}{H}\right)}{\partial y} = -gH\frac{\partial \eta}{\partial x} + 2\frac{\partial}{\partial x}\left(\mu H\frac{\partial u}{\partial x}\right) + \frac{\partial}{\partial y}\left[\mu H(\frac{\partial v}{\partial x} + \frac{\partial u}{\partial y}\right] - FU$$

$$\frac{\partial U}{\partial t} + \frac{\partial \left(\frac{UV}{H}\right)}{\partial x} + \frac{\partial \left(\frac{V^2}{H}\right)}{\partial y} = -gH\frac{\partial \eta}{\partial x} + \frac{\partial}{\partial x}\left[\mu H(\frac{\partial v}{\partial x} + \frac{\partial u}{\partial y}\right] + 2\frac{\partial}{\partial y}\left(\mu H\frac{\partial v}{\partial y}\right) + -FU$$

with,

$$H = d + \eta; U = Hu; V = Hv$$

$$F = gH^{-\frac{7}{3}}n^2\left(U^2 + V^2\right)^{\frac{1}{2}}$$

$$\mu = \mu_0 + \mu_\tau$$

where:

d	water depth
η	free surface elevation
g	acceleration due to gravity
n	Manning coefficient
u, v	depth averaged velocities
μ_0, μ_τ	kinematic and eddy viscosity

In a large, shallow embankment, the horizontal viscosity term is much smaller than the vertical viscosity term, and is often neglected [77]. The vertical term is then parameterized by the bottom friction term similar to the classical boundary layer theory [84].

We will now try to understand the underlying physical processes of the Coanda effect in jet water interactions in shallow waters based on the works of Lalli et al. [85]. The authors had carried out numerical and experimental studies on such interactions in shallow waters, and particularly in two Italian harbors, namely the Pescera Channel harbor and the Latina harbor.

In numerical simulations, the authors made several simplifications. They linearized shallow water equations with respect to the free surface elevation, and assumed the elevation to be low, which implied that the Froude number, F_r was less than 0.4, and the flow field everywhere under investigation was subcritical. They further neglected the stratification effects due to salinity and temperature. The authors, for simplicity, used algebraic model [85, 86] rather than k–ε model [87] for turbulence closure which was given by:

$$\mu = CH\mu_\tau$$

where C is a parameter that includes three-dimensional effects.

4.8.1 Experimental Setup and Test Configurations

Figure 4.57 shows the experimental setup that consisted of a flat-bed rectangular tank with a horizontal, rectangular, cross-sectioned, free surface channel placed normal to the tank to represent the inlet of river. The outlet section was placed opposite the channel mouth. Measurements were carried out by particle image velocimetry.

Four experimental configurations were investigated in this study. They were:

A. rectilinear lateral wall, parallel to the jet axis, placed at a distance h from river mouth
B. simple-shaped channel-harbor layout, with a breakwater and an L-shaped jetty
C. Pescara Harbor (1:1000 scale), and
D. Latina harbor proposed design (1:700 scale)

We will only look at the results of configurations A and D as being sufficient to highlight the main features.

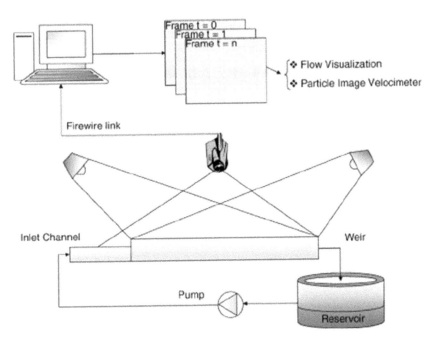

FIGURE 4.57
Experimental setup used in shallow water investigation. (After: [85].)

4.8.2 Experimental Results

4.8.2.1 Jet–Wall Interaction for Configuration A

Figure 4.58 (a) and Figure 4.58 (b) show the experimental and numerical vector plots of time averaged jet–wall interaction in shallow waters for Configuration A. The flow attachment depicting the Coanda effect resembles the flow features of a backward facing step.

The effect of bottom friction on the Coanda effect using configuration A can be seen in Figures 4.59 (a) and (b). These are the numerical solutions expressed in streamlines for the same flow except that the Manning roughness in 4.59 (a) was five times larger than in 4.59 (b).

4.8.2.2 Jet–Wall Interaction for Configuration D

Figures 4.60 (a)–(f) show the experimental and numerical flow patterns at different time periods for the Latina Harbor configuration. The figures on the left show the experimental flow visualization results while the right-hand

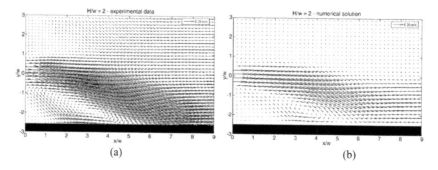

FIGURE 4.58
(a)–(b): Jet-wall interaction in shallow water for Configuration A. (a) Experimental data; (b) Numerical solution. (After: [84].)

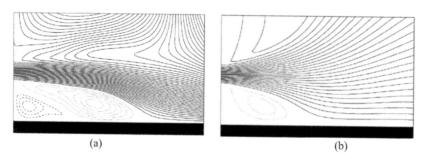

FIGURE 4.59
(a)–(b): Effect of bottom friction. (After: [85].)

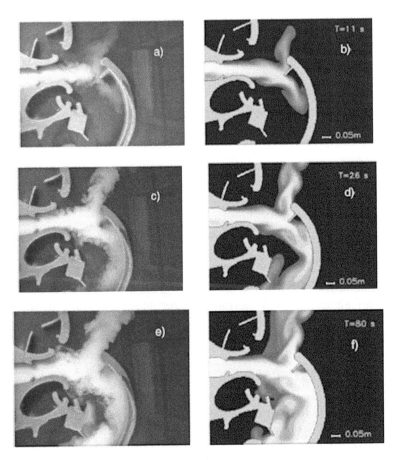

FIGURE 4.60
(a)–(f): Numerical and experimental result comparisons of river water diffusion with progression of time. (After: [85].)

side shows the numerical results. The diffusion of water in the Mascarello River as it interacts with the Latina Harbor structures over time can be seen in these figures.

Despite large simplifications of the numerical and experimental models, the agreement between the experimental and numerical results in terms of capturing the flow attributes were quite satisfactory.

The following remarks specific to the impacts of the Coanda effect can be made from the results of this study:

- The Coanda effect is essentially an inviscid flow phenomenon driven by convective terms and free surface gradients. In other words, the forcing terms that initiate the Coanda effect are the free surface gradients related to the vortex generated between the jet and the wall aided by convective terms that produces flow contraction.

- High bottom friction inhibits the Coanda effect. If the ratio of the horizontal to vertical scale is sufficiently large, the Coanda effect may not take place at all.

- The Coanda effect significantly modifies the flow patterns in jet–flow interactions by affecting the diffusion process and the spreading of river waters along the coast. The authors suggested that construction of a breakwater in front of a river mouth is not suitable because it hinders the free spreading flow of river waters, resulting in undesirable stagnation of polluted waters.

4.9 Building Ventilation

Ventilation, or the introduction of air into an enclosed space [88–91] to control indoor air quality and thermal comfort, is essential in human living. Various strategies involving renewable and conventional sources of energy have been employed [92–94] for the purpose.

4.9.1 HVAC Ceiling Diffuser

Heating ventilation and air-conditioning (HVAC) is one of the commonly used systems in building ventilation [95]. However, to maintain a habitable indoor environment for occupants within a building, the system has to operate under varying load conditions, and provide maximum heating in winter and maximum cooling in summer, as well as meet the requirements of more moderate days. These systems, however, are energy intensive and costly to install and operate.

To create a comfortable condition, the air supplied from the HVAC draws in the air of the room and mixing takes place [96]. A schematic of such mixing is shown in Figure 4.61. The diffuser of the HVAC is used to supply the air, generally from the ceiling of the room. The diffuser generates a controlled high-velocity jet, creating the Coanda effect, and the air leaves the diffuser, adhering to the ceiling [97] as shown in Figure 4.62.

The performance of the diffuser may be determined by the length of the jet movement on the room ceiling. Figure 4.63 shows the ideal movement of the jet that reaches each corner of the ceiling [98] before flowing downwards, inducting air of the room into mixing.

4.9.2 Free Plane Jet Ventilation

Ventilation by a free plane jet with the Coanda effect has been studied by van Hooff et al. [99]. The authors conducted particle image velocimetry (PIV)

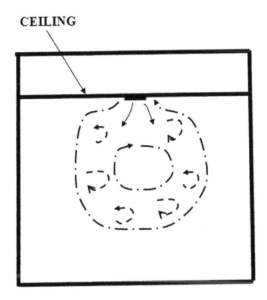

FIGURE 4.61
Air induction by supply air and mixing within a room.

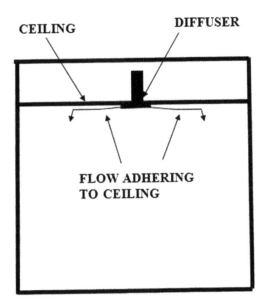

FIGURE 4.62
Coanda effect produced by supply air jets.

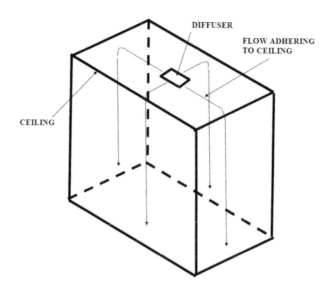

FIGURE 4.63
An isometric view of the ideal movement of ceiling supply jet movement to each corner of the room.

measurements and analysis of forced mixing ventilation. A reduced scale experimental setup, which is shown in Figure 4.64 (a)–(b), was used for their investigation.

Based on the PIV measurements, time-averaged velocities, vorticity, and turbulent intensities, the authors found that the Coanda effect caused the free plane jet to transform in the wall at a distance just downstream of the inlet. The Coanda effect was found to deflect the jet toward the top surface. The development of the Coanda effect at different Reynolds numbers is shown in Figure 4.65.

4.9.3 Double-Glazed Facades

Double-glazed façades are increasingly seen as useful means to compliment the building ventilation process. Apart from increasing the aesthetic appeal of modern architecture, such façades also serve to shield a building from noise and wind loads. These façades, however, allow ultraviolet rays of the sun to enter the building. Although ultraviolet rays help in the formation of vitamin D necessary for strengthening the bone, prolonged exposure to this ray may cause burns and skin cancer.

Under varying climatic conditions, the thermal performance of a double-glazed façade is generally not satisfactory and may require extra efforts to maintain livable conditions inside a building. To reduce the solar load gain due to the installation of double glazed faces, for example, free and forced convection methods or a combination of both methods may be required [100–102].

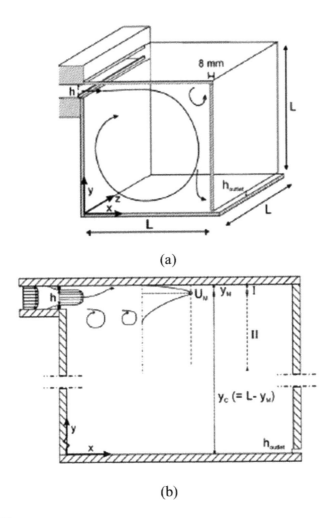

(a)

(b)

FIGURE 4.64

(a)–(b): Reduced scale PIV measurement set up: (a) three-dimensional view of the test section and coordinate system used; (b) two-dimensional representation of the plane jet for the inner and outer regions I and II. (After: van Hooff, et al., 2012 [99].)

Free convection that relies on buoyancy is less effective and forced convection using mechanical means appears to produce better results. Fans are the most commonly used in forced convection ventilation in a double-glazed façade. However, Valentín et al. [103] have pointed out several disadvantages of using fans, such as the costs of an electrical supply for the motor and the construction of a solid support in the lateral side of the façade for their installation, as well as the issues of vibration and noise, and additional maintenance costs.

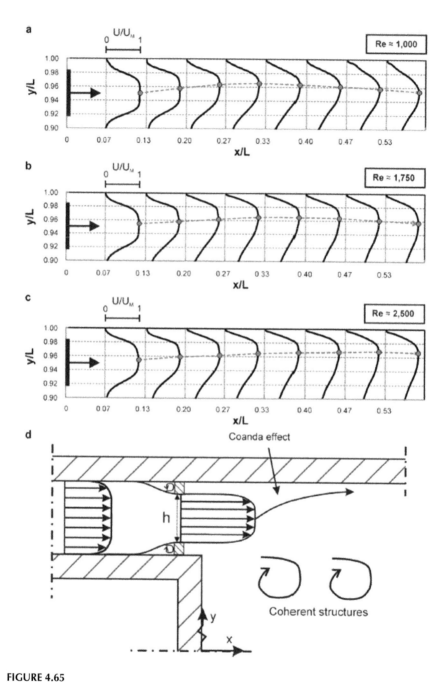

FIGURE 4.65
(a)–(d): Jet development and the Coanda effect at various Reynolds numbers in mixed forced ventilation. (After: van Hooff, et al., 2012 [99].)

4.9.3.1 Coanda Nozzles

In the same study, Coanda nozzles are claimed to overcome these problems; they carried out some feasibility studies on them using computational fluid dynamics. We will briefly look at the results from their studies on the performance of the Coanda nozzle. An isometric view of the Coanda nozzle is shown in Figure 4.66.

In the proposed nozzle, a thin layer of fluid is ejected at a high speed from a gap near the throat on the circumference. This creates suction that allows the ejected fluid to move while remaining attached to the wall and entraining atmospheric air through the nozzle.

4.9.3.2 Performance Evaluation

The authors resorted to numerical investigation to evaluate the performance of the proposed Coanda nozzle. The geometric model used in their study was constructed following the actuator optimization guidelines of Kim et al. [104]. The schematic of the vertical section of the double-glazed façade and the radial section of the Coanda nozzle are given in Figure 4.67 (a) and (b).

The results obtained were compared with the results of Guardo et al. [105, 106]. The Coanda nozzle ventilation results suggested that the solar load gain was reduced significantly over vertical ventilation and produced high volumetric flow rates. A recirculating flow pattern was observed through the lower zone of the façade indicating heat removal mechanism. These findings led the authors to conclude that the Coanda nozzle was a feasible alternative to the conventional fans for forced convection ventilation in double-glazed façades.

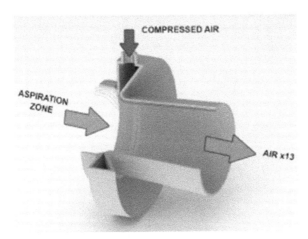

FIGURE 4.66
An isometric view of a Coanda nozzle. (After: Valentin, et al., 2013 [103].)

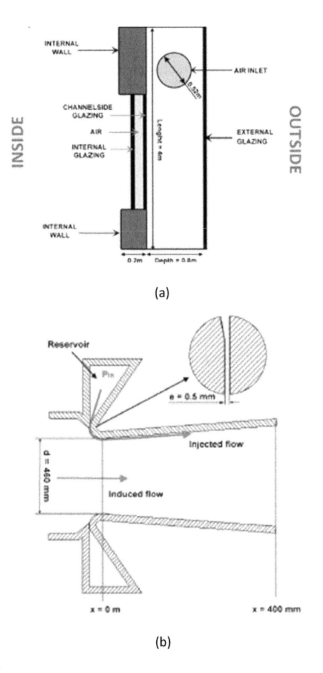

(a)

(b)

FIGURE 4.67
(a)–(b): Details of the geometric model used in the numerical study: (a) vertical section of the double-glazed façade; (b) radial section of Coanda nozzle. (After: Valentin, et al., 2013 [103].)

4.10 Drinkable Water Systems

Water is a basic requirement of all living beings. Human civilizations and agriculture have grown around water supplied by rivers and their branches. With continued population growth, the need for improved water supply systems for drinkable water has also grown. Unfortunately, however, nearly half the population of the world living in the rural areas [107] of developing countries does not have access to drinkable water.

In many parts of the developing world, small feed water systems are generally used. These systems require the removal of sediment from the water before they become drinkable. Proper removal of sediment from the water of the river systems is often a difficult task particularly for the fact that the sediment load varies at different times of the year due to seasonal changes.

To illustrate the impact of sedimentation on the feed water system, which can affect millions of lives, let us briefly consider some statistics related to the Mekong River, one of the longest rivers in the world that runs through several heavily populated countries of Southeast Asia.

Figure 4.68 shows a map of the Mekong River [108], and a view of the river as it passes through Luang Prabang in Laos [109] can be seen in Figure 4.69.

The Mekong River is 4,350 km long [110] and it drains an area of 795,000 km^2, discharging 475 km^3 of water annually [111]. According to the Mekong River Commission of 2009 [112], the river has a peak flow of about 12,000 m^3/s with an average flow of about 3,500 m^3/s annually at Luang Prabang in Laos. This translates into a peak-to-average-daily-discharge ratio of 3.4. This ratio increases rapidly as the basin area becomes smaller [110]. This ratio is important because it shows how in rainy seasons, it can result in watershed erosion and increase the sediment load.

The quality of water is greatly affected by the sediment present in the river water, which is often the main source of water. The performance of water treatment plants is also adversely affected by the sediments. Although various attempts have been made to provide drinkable water to rural communities of the developing world, the cost of installation, maintenance, as well as a lack of resources have made the task of providing drinkable water in those communities much harder.

In this section, we will look at how the Coanda effect can contribute to improving water quality and water supply systems, particularly for the rural poor of developing countries. The device that can be used is the Coanda effect screen, also referred to as a wedge wire screen.

4.10.1 Working Principle of the Coanda Screen

Coanda effect screens have evolved from screen designs that have been used for a long time in the mining and wastewater treatment industries to separate liquids from solids over an inclined plane of standard wedge-wire

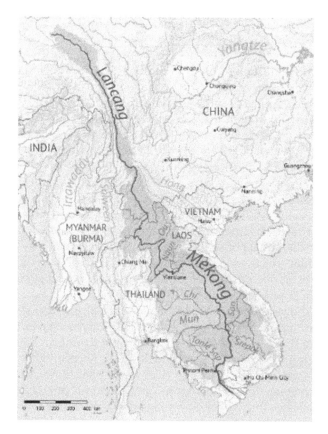

FIGURE 4.68
Map of the Mekong River Basin. (After: Shannon1 [108].)

screen panels in which the top surface of each wire is parallel to the plane of the complete screen.

The through-flow is driven by the effective hydrostatic pressure created by the water column height over the gap between any two parallel wires and the thin layer of water sheared off at the bottom of the section.

The mining and wastewater wire panels generally operate on gravity action only. The Coanda effect screen panels are intended to gradually turn the flow in such a way that the flow progresses without skipping a single wire. Thus, each individual wire of the Coanda effect screen panel is tilted by a few angles (3–6°) downstream in relation to the previous upstream panel while at the same time placed in such a way as to produce very small openings (less than 1 mm). At typical velocities and screen openings, this deflection is very slight. With tilted wires, the shearing action is also enhanced by the attachment of the flow, which is the Coanda effect. The shearing effect

FIGURE 4.69
A view of the Mekong River as it passes through Luang Prabang, Laos. (After: [109].)

and the orifice-flow behavior influence the design parameters of the Coanda-effect screen.

The screen properties of slot width, wire width, and wire tilt angle can significantly affect screen capacity. Slot width and wire width both affect screen porosity, which affects the amount of orifice-type flow through the screen surface. Wire tilt angle also affects the shearing of flow through the screen.

Several studies [112–114] have been conducted to relate the orifice discharge coefficient to the Froude number between parallel slot flows. They used one-dimensional flow with a prescribed distribution of hydrostatic pressure as well as different free surface models demonstrating the viability of the numerical modelling approach. The results obtained, however, were only of indicative nature since the flows modelled were through parallel slots of rectangular bars that were substantially different in configuration from the Coanda effect screen slots.

The essential features of a Coanda effect screen and installation can be seen in Figure 4.70 [115]. The screen is installed on the downstream face of an overflow weir. Flow passes over the crest of the weir, across a solid acceleration plate, and then across the screen panel. The screen is, generally, constructed of wedge-wire where the wires are placed horizontally but normal to the flow across the screen.

Wahl [116] showed that the through-flow discharge through the tilted wires could be described in terms of the orifice equation. This equation was expressed as a function of two coefficients; the first accounted for the

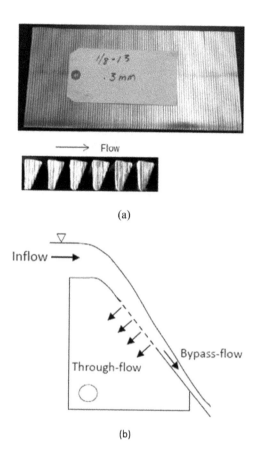

FIGURE 4.70

(a)–(b): Features of a Coanda effect screen: (a) photographs of a typical Coanda effect screen in plan and longitudinal section views; (b) installation of the Coanda effect screen on the downstream face of an overflow weir. (After: May, 2015 [115].)

reduction and contraction in velocity through the slot, while the second captured the effects of the Froude number, geometry of the wire, and the angle of incidence of the flow. In this analysis, the specific energy term from the depth of the flow was used as the driving force for the flow with a Froude number that ranged between 2 and 30. The study also found that although the screen curvature and air entrainment affected the performance of the Coanda effect screen, the hydraulic friction of the screen had little or no effect.

4.10.2 Optimum Performance Screens

For optimum performance of the Coanda effect screen, upstream flow should accelerate smoothly across downstream across the face of the screen.

The relationship between the inclination of the screen and the drop height, as well as the shape of the accelerator that is meant to facilitate the smooth transition, therefore, become very important. These aspects have been studied by Wahl [115, 116], who found that the ideal accelerator plate profile should follow the trajectory of a free falling jet under the action of gravity, in other words an ogee shape. The exact ideal ogee shape, however, depends on the flow discharge, and, to a lesser extent on the height and velocity of the upstream flow. A qualitative description of the ogee shapes for different discharge values is shown in Figure 4.71.

4.10.3 Design Parameters

In a later work, Wahl [117] using equations and information contained in Reference [118], developed various curves to determine the drop height, screen inclination, or design discharge of an ogee crest accelerator plate when any two of these three parameters was known in the design of any Coanda effect screen. A qualitative description of the relationship for the drop height and flow discharge based on Wahl's work [116] for various Coanda effect screens is shown in Figure 4.72.

Wahl's work was conducted using clean water on Coanda effect screens. Although Kamanbedast [119] studied sediment removal at the intakes of small hydropower installations, he used parallel rectangular bars, and as such not Coanda effect screens. May [115] also studied the effectiveness of Coanda effect screens in removing sediment. His work will now be discussed.

May [115] conducted several laboratory experiments to determine the effectiveness of Coanda effect screens in removing sediment. Initially, he used

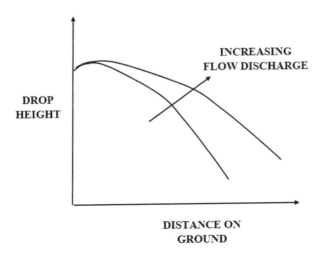

FIGURE 4.71
Qualitative description of Ogee crest profiles for varying discharge values.

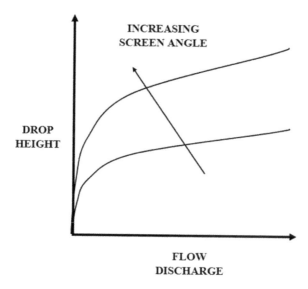

FIGURE 4.72
Qualitative description of relationship between drop height and flow discharge on various screen angles.

a continuous flow apparatus with clean water to validate the performance of Coanda effect screens. His results were found to be compatible with the findings of Wahl. The apparatus was, thereafter, modified to conduct experiments in batch flow mode for sediment exclusion. The schematic diagram of the modified apparatus is shown in Figure 4.73.

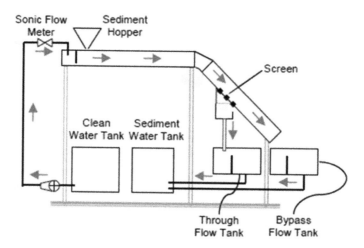

FIGURE 4.73
Schematic diagram of the apparatus for a batch flow experiment. (After: May, 2015 [115].)

4.10.4 Experiment

4.10.4.1 Laboratory test

The study found, as expected, that with time the gap between the individual wires tended to clog with sediments. This had the effect of reducing the through-flow depending on the geometry of the screens used. Figure 4.74 shows this behavior.

It is interesting to note, however, that the clogged materials were removed periodically by what can be described as something akin to a self-cleaning process. This aspect was also observed in the experiments. The increased shear stress due to increased depth and local velocity was probably the factors behind this behavior.

The results of sediment exclusion experiments performed under laboratory conditions are, however, difficult to implement in a real environment because of the difficulties of knowing the sediment particle size distribution

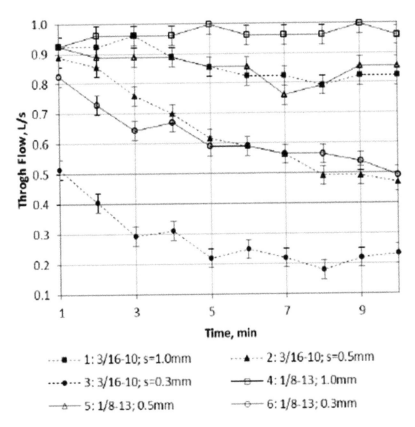

FIGURE 4.74

Sediment mixed flow through a Coanda effect screen as a function of time. (After: May, 2015 [115].)

and concentration that vary under the local topology, geology, and hydrology conditions. The sediment load also changes at different times of the year.

4.10.4.2 Field Trial Test

May [115], therefore, built a prototype for field testing similar to the types of gravity water systems used in the rural areas of many developing countries. Particular attention was paid to the cost and time in building them to assess their economic viability. Figure 4.75 shows the Field Prototype that was built.

The structure was built for a maximum through-flow of 1.26 liters/second, with a spillway slope angle of 45°, a drop height of 66 mm, and channel width of 15 cm. The area of the screen was 15 cm × 30 cm. A provision was included to decrease the screen length to accommodate different through-flow rates. The top two pictures in Figure 4.75 show a pipe on the left of the structure intended to collect and transport the through-flow downstream. Debris is seen to have been collected on the bottom-left picture of Figure 4.75, which is eventually removed by the flow.

Although it was difficult to establish a definitive performance correlation between the wire width, w, and tilt angle, Φ, screens with thinner wire seemed to be better at removing sediment, but this came at the cost of lower

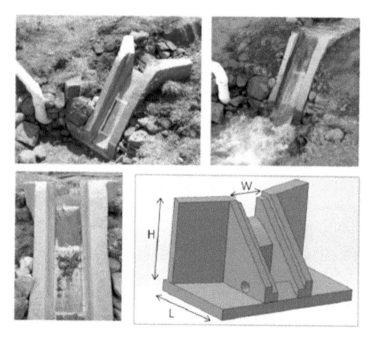

FIGURE 4.75
Field prototype of a Coanda effect screen used. (After: May, 2015 [115].)

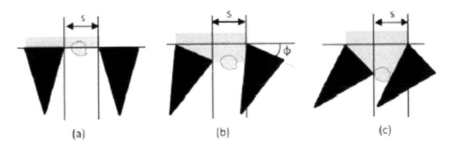

FIGURE 4.76
(a)–(c): Effect of tilt angle variation. [115].

through-flow rate. May conjectured that the tilt angle was the more important of the two parameters and used the schematic diagram shown in Figure 4.76 (a)–(c) to argue the case. The diagram shows the increase from 0° in frame (a) to (c) with progressively increasingly values of Φ. As tilt angle increases, the through-flow also increases with a greater possibility of trapping sediment particles on the Coanda effect screen.

4.11 Concluding Remarks

We had a glimpse of the diverse application possibilities of the Coanda effect and how they can influence our lives, from automation of industrial processes to healthy living with drinkable water systems. Still there are other works that show vast opportunities exist for Coanda effect applications in areas such as vacuum cleaning; discharging foul air or dust from chicken houses away from residential homes [120]; high performing automotive vehicles [121]; heat exchangers for racing cars [122]; fluidic actuators for control systems [123]; flow meters for hydrocarbon rate measurement in the oil and gas industries; noise reduction of feedback control systems [124]; and so forth. Coanda effect research is providing promising solutions.

References

1. Iguchi, M., Terauchi, Y., and Yokoya, S., (1998), Effect of cross-flow on the frequency of bubble formation from a single-hole nozzle, *Metallurgical and Materials Transactions B*, vol. 29, no. 6, pp. 1219–1225.

2. Iguchi, M., and Sasaki, K., (2006), Prediction of argon gas attachment to sliding gate in immersion nozzle, *ISIJ International*, vol. 46, no. 8, pp. 1264–1266.
3. Maeda, D., and Iguchi, M., (2002), Steady and periodical bubble attachment to the sliding gate in immersion nozzle, *ISIJ International*, vol. 42, no. 11, pp. 1196–1202.
4. Uemura, T., Iguchi, M., and Ueda, Y. (2018), Miscellaneous applications, In Masato Wakayama (editor-in-Chief), *Flow Visualization in Materials Processing. Mathematics for Industry*, vol. 27, pp. 89–168 Springer, Tokyo.
5. Sonoyama, N., and Iguchi, M., (2002), Bubble formation and detachment on nonwetted surfaces, *Metallurgical and Materials Transactions B*, vol. 33, no. 2, pp. 155–162.
6. Iguchi, M., and Ilegbusi, O.J., (2010), Interfacial phenomena, In *Modeling Multiphase Materials Process*, pp. 89–168, Springer, New York.
7. Pera, L., and Gebhart, B., (1975), Laminar plume interactions, *Journal of Fluid Mechanics*, vol. 68, pp. 259–271.
8. Freire, A.P.S., Miranda, D.D.E., Luz, L.M.S., and Franca, G.F.M., (2002), Bubble plumes and the Coanda effect, *International Journal of Multiphase Flow*, vol. 28, pp. 1293–1310.
9. Herringe, R.A., and Davis, M.R., (1974), Detection of instantaneous phase changes in gas–liquid mixtures, *Journal of Physics E: Science Instrumentation*, vol. 7, pp. 807–812.
10. Kobus, H.E., (1968). Analysis of the flow induced by air-bubble systems, in *Proceedings of 11th Coastal Engineering Conference London*, vol. 2, pp. 1016–1031.
11. Ditmars, J.D., and Cederwall, K., (1974). Analysis of air-bubble plumes, in *Proceedings of Coastal Engineering Conference*, pp. 2209–2226.
12. Iguchi, M., and Sasaki, K., (1999), Coanda rising effect on liquid flow near the side wall of a characteristics cylindrical in a vessel bubbling jet, *ISIJ International*, vol. 39, no. 3, pp. 213–218.
13. Iguchi, M., Sasaki, K., Nakajima, K., and Kawabata, H., (1998), Coanda effect near the side on bubble wall of a characteristics in a cylindrical vessel, *ISIJ International*, vol. 38, no. 12, pp. 1297–1303.
14. Inoue, T., Iguchi, M., and Mizuno, Y., (2000), Separation of gas and liquid based on wettability difference of circular pipes, in *Proceedings of 1st Japan Society, Multiphase Flow Conference*, pp. 241–242.
15. Mizuno, Y., Inoue, T., and Iguchi, M., (2001), *Proceedings of the 3rd Pacific Symposium on Flow Visualization and Image Processing*, Outrigger Wailea Resort, Pacific Center of Thermal-Fluid Engineering, Hawaii, USA.
16. Mizuno, Y., and Iguchi, M. (2001), Behavior of bubbles attaching to and detaching from solid body of poor wettability, *ISIJ International*, vol. 41(Supplement), pp. S56–S59.
17. Mizuno, Y., Yamashita, T., Iguchi, M., and Morita, Z., (2001), *Proceedings of 2nd International Congress on Science and Technology of Steelmaking*, University of Wales, Swansea, p. 513.
18. Yamashita, T., Mizuno, Y., and Iguchi, M., (2001), *Proceedings of JSMF Annual Meeting 2001*, Japan Society of Multiphase Flow, Osaka, p. 17.
19. Ogawa, Y., and Tokumitsu, N., (1990), Observation of slag foaming by X-ray fluoroscopy, in *6th International Iron and Steel Congress*, ISIJ, Tokyo, vol. 1, pp. 147–152.

20. Iguchi, M., Takeuchi, H., and Morita, Z., (1991), *The Flow Field in Air-Water Vertical Bubbling Jets in a Cylindrical Vessel, I-STAGE home/ISIJ International*, vol. 31, no. 3, p. 246.
21. Sasaki, K., Iguchi, M., Ishi, T., and Yokoya, S., (2002), The Coanda effect on bubbling jet behind a horizontally placed circular cylinder, *ISIJ International*, vol. 42, no. 11, pp. 1196–1202.
22. Iguchi, M., Ueda, H., and Uemur, T., (1995), Bubble and liquid flow characteristics in a vertical bubbling jet, *International Journal of Multi-phase Flow*, vol. 21, no. 5, pp. 861–873.
23. Iguchi, M., Nakamura, K., and Tsujino, R., (1998), Mixing time and fluid flow phenomena in liquids of varying kinematic viscosities agitated by bottom gas injection, *Metallurgical and Materials Transactions B*, vol. 29, no. 3, pp. 569–575.
24. Oliveira, J.F.G., Silva, E.J., Guo, C., and Hashimoto, F., (2009), Industrial challenges in grinding, *Annals of the CIRP*, vol. 58, no. 2, pp. 663–680.
25. Malkin, S., Guo, C., (2007), Thermal analysis of grinding. *Annals of the CIRP*, vol. 56, pp. 760–782.
26. Shibata., J., Goto, T., Yamamoto, M., and Tsuwa, H., (1952), Characteristics of air flow around a grinding wheel and their availability for assessing the wheel wear, *Annals of the CIRP*, vol. 31, no. 1, pp. 233–238.
27. Brinksmeier, E., Heinzel, C., and Wittmann, M., (1999), Friction, cooling and lubrication in grinding, *Annals of the CIRP*, vol. 48, no.2, pp. 581–598.
28. Irani, R.A., Bauer, R.J., and Warkentin, A., (2005), A review of cutting fluid application in the grinding process, *International Journal of Machine Tools & Manufacture*, vol. 45, pp. 1696–1705.
29. Webster, J., Brinksmeier, E., Heinzel, C., Wittmann, M., and Thoens, K., (2002), Assessment of grinding fluid effectiveness in continuous-dress creep feed grinding, *Annals of the CIRP*, vol. 51, no. 1, pp. 235–2400.
30. Yoshimi, T., Oishi, S., Okubo, S., and Morita, H., (2010), Development of minimized coolant supply technology in grinding, *JTEKT Engineering Journal*, edition no. 1007E, pp. 54–59.
31. Alberdi, R., Sanchez, J.A., Pombo, I., Ortega, N., Izquierdo, B., Plaza, S., and Barrenetxea, D., (2011), Strategies for optimal use of fluids in grinding, *International Journal of Machine Tools and Manufacture*, vol. 5, pp. 491–499.
32. Ninomiya, S., Suzuki, K., Uematsu, T., Iwai, M., and Tanaka, K., (2004), Effect of a small gap nozzle facing to grinding wheel in slit grinding, *Key Engineering Materials*, 257, 258, pp. 327–332.
33. Morgan, M.N., Jackson, A.R., Wu, H., Baines-Jones, V., Batako, A., and Rowe, W.B., (2008), Optimization of fluid application in grinding, *Annals of the CIRP*, vol. 57, no.1, pp. 363–366.
34. Lopez-Arraiza, A., Castillo, G., Hom, N., Dhakal, H.N., and Alberdi, R., (2013), High performance composite nozzle for the improvement of cooling in grinding machine tools, *Composites Part B*, vol. 54, pp. 313–318.
35. Hosokawa, A., Tokunaga, K., Ueda, T., Kiwata, T., and Koyano, T., (2016), Drastic reduction of grinding fluid flow in cylindrical plunge grinding by means of contact-type flexible brush-nozzle, *CIRP Annals – Manufacturing Technology*, vol. 65, pp. 317–320.
36. Ferrari, J., Lior, N., and Slycke, J. (2003), An evaluation of gas quenching of steel rings by multiple-jet impingement, *Journal of Material Process Technology*, vol. 136, pp. 190–201.

37. Martin, H. (1977), Heat and mass transfer between impinging gas jets and solid surfaces, *Advances in Heat Transfer*, vol. 13, pp. 1–60.
38. Jambunathan, K., Lai, E., Moss, M.A., and Button, B.L., (1992). A review of heat transfer data for single circular jet impingement, *International Journal of Heat Fluid Flow*, vol. 13, pp. 106–115.
39. Donaldson, C.D., and Snedeker, R.S. (1971), A study of free jet impingement. Part 1. Mean properties of free and impinging jets, *Journal of Fluid Mechanics*, vol. 45, pp. 281–319.
40. Viskanta, R. (1993), Heat transfer to impinging isothermal gas and flame jets. *Experimental Thermal Fluid Science*, vol. 6, pp. 111–134.
41. Zuckerman, N., and Lior, N., (2006), Jet impingement heat transfer: physics, correlations, and numerical modeling, *Advances in Heat Transfer*, vol. 39, pp. 565–631.
42. Kramer, C., Gerhardt, H.J., and Knoch, M., (1984), Applications of jet flows in industrial flow circuits, *Journal of Wind Engineering and Industrial Aerodynamics*, vol. 16, no. 2–3, pp. 173–188.
43. By Richard Greenhill and Hugo Elias (myself) of the Shadow Robot Company – http://www.shadowrobot.com/media/pictures.shtml, CC BY-SA 3.0; https://commons.wikimedia.org/w/index.php?curid=5081565
44. Shi, K., and Li, X., (2016), Optimization of outer diameter of Bernoulli gripper, *Experimental Thermal and Fluid Science*, vol. 77, pp. 284–294.
45. Gualtiero, F., Santochi, M., Gino, D., Dini, Tracht, K., Scholz-Reiter, B., Fleischer, J., Kristoffer, T., Lien, T., Seliger, G., Reinhart, G., Franke, J., Hansen, H.N., and Verl, A., (2014), Grasping devices and methods in automated production processes, *CIRP Annals*, vol. 63, no. 2, pp. 679–701.
46. Dinn, G., Fantoni, G., and Failli, F., (2009), Grasping leather plies by Bernoulli grippers, *CIRP Annals*, vol. 58, no. 1, pp. 21–24.
47. Seliger, G., Szimmat, F., Niemeyer, J., and Stephan, J., (2003), Automated handling of nonrigid parts, *Annals of the CIRP*, vol. 52, no.1, pp. 21–24.
48. Dougeri, Z., and Fahantidis, N., (2002), Picking up flexible pieces out of a bundle, *IEEE Robotics and Automation Magazine*, vol. 9, no. 2, pp. 9–19.
49. Ameri, M., and Dybbs, A., (1993), Theoretical modeling of Coanda ejectors, American Society of Mechanical Engineers, Fluid Engineering Division (Publication) FED, Fluid Machinery, vol. 163, pp. 43–48.
50. Lien, T.K., and Davis, P.G.G, (2008), A novel gripper for limp materials based on lateral Coanda ejectors, *CIRP Annals – Manufacturing Technology*, vol. 57, pp. 33–36.
51. Jain, A.K., Briegleb, B.P., Mininswaner, K., and Wubbles, D.J., (2000), Radiative forcings and global warming potentials of 39 greenhouse gases, *Journal of Geophysical Research: Atmospheres*, vol. 105, no. D16, pp. 20773–20790.
52. By Josua Doubek – Own work, https://commons.wikimedia.org/wiki/File:North_Dakota_Flaring_of_Gas.JPG
53. Figure: Flaring of associated gas from an oil well site in Nigeria, author Chebyshev1983, https://en.wikipedia.org/wiki/File:Niger_Delta_Gas-Flares.jpg
54. By Terryjoyce – Own work, CC BY-SA 3.0, https://commons.wikimedia.org/w/index.php?curid=5667339
55. Photograph by Verodrig – Own work, CC0, https://commons.wikimedia.org/wiki/File:First_gas_from_the_Oselvar_module_on_the_Ula_platform_on_April_14th,_2012.jpg

56. CBC News (online), (September 17, 2013), 7,500 songbirds killed at Canaport gas plant in Saint John. www.cbc.ca/news/canada/new-brunswick/7-500-song-birds-killed-at-canaport-gas-plant-in-saint-john-1.1857615

57. Wiese, F.K., Montevecchi, W.A., Davoren, G., and Lake, G., (2001), Seabirds at risk around offshore oil platforms in the North-west Atlantic, *Marine Pollution Bulletin*, vol. 42, no. 12, pp. 1285–1290.

58. https://www.premiumtimesng.com/business/business-news/230059-nige ria-lost-850-million-gas-flaring-2015-official.html, April 30, 2017.

59. World Bank, Brochure (2011), Global Gas Flaring Reduction Partnership (GGFR), World Bank, October.

60. United States Energy Information Administration, (2012), Annual Energy Review, Table 6.7 Natural Gas Wellhead, Citygate, and Imports Prices, 1949–2011, September.

61. Watts, P., Advances in Flare Technology, Brochure of Kaldair Ltd, https://ww w.scribd.com/document/35366210/Advanced-in-Flare-Technolofy-Kaldair

62. Wilkins, J., Witheridge, R.E., Desty, D.H., Mason, J.T.M., and Newby, N., (1997), *Proceedings of the Ninth Annual Offshore Technology Conference*, Houston, TX, pp. 123–130 (The design, development and performance of the Indair and Mardair flares)

63. Desty, D.H., Boden, J.C., and Witheridge, R.E., (1978), The origination, development and application of novel premixed fare burners employing the Coanda effect, in *Proceedings of the 74th National Meeting of the American Institute of Chemical Engineering*, Philadelphia, PA.

64. Desty, D.H., (1983), No smoke with fire, *Proceedings of the Institution of Mechanical Engineers*, vol. 197A, pp. 159–170.

65. Jenkins, B.G., Moles, F.D., Desty, D.H., Boden, J.C., and Pratley, G., (1980), The aerodynamic modelling of fares, in Energy Technology Conference and Exhibition, New Orleans, LA, ASME Paper 80-PET–86.

66. Fricker, N., Cullender, R.H., Brien, K.O., and Sutton, J.A., (1986), Coanda jet pumps-facts and fallacies, in *Proceedings of the International Gas Research Conference*, Toronto, Canada (Cramer, T.L., editor) Government Institutes Inc., Rockville, MD, pp. 989–1003.

67. Carpenter, P.W., and Green, P.N., (1997), The aeroacoustics and aerodynamics of high-speed Coanda devices, part 1: conventional arrangement of exit nozzle and surface, *Journal of Sound and Vibration*, vol. 208, no. 5, pp. 777–801.

68. Carpenter, P.W., and Green, P.N., (1997), The aeroacoustics and aerodynamics of high-speed Coanda devices, part 2: effects of modifications for flow control and noise reduction, *Journal of Sound and Vibration*, vol. 208, no. 5, pp. 803–822.

69. Gregory-Smith, D.G., and Senior, P., (1994), The effects of base steps and axi-symmetry on supersonic jets over Coanda surfaces, *International Journal of Heat and Fluid Flow*, vol. 15, pp. 291–298.

70. Gaydon, A.G., and Wolfhard, H.G., (1979), *Flames*, 4th edition, Chapman and Hall, New York, p. 169.

71. Gil, Y.S., Jung, H.S., and Chung, S.H., (1998), Premixed flame stabilization in an axisymmetric curved-wall jet, *Combustion and Flame*, vol. 11, pp. 348–357.

72. Lewis, B., and von Elbe, G., (1987), Combustion, flames, and explosion of gases, 3rd edition, Academic, Orlando, pp. 1–739.

73. Photograph by Rodney Haywood, public domain, https://commons.wikimedia .org/wiki/File:Sydney_Harbour_Bridge_from_the_air.JPG

74. Public domain, https://upload.wikimedia.org/wikipedia/commons/5/5a/Ter racinaPortoCanale.jpg

75. https://commons.wikimedia.org/wiki/File:Rip_current_warning_signs_at Beach

76. Wind, H.G., and Vreugdenhil, C.B., (1986), Rip-current generation near structures, *Journal of Fluid Mechanics*, vol. 171, pp. 459–476.

77. Casulli, V., and Cheng, R.T., (1992), Semi-implicit finite difference methods for three dimensional shallow water flow, *International Journal of Numerical Methods in Fluids*, vol. 15, pp. 629–648.

78. Guillou, S., and Nguyen, K.D., (1999), An improved technique for solving two-dimensional shallow water problems, *International Journal of Numerical Methods in Fluids*, vol. 29, pp. 465–483.

79. Haeuser, J., Paap, H.G., Eppel, D., and Mueller, A., (1985), Solution of shallow water equations for complex flow domains via boundary-fitted co-ordinates, *International Journal of Numerical Methods in Fluids*, vol. 5, pp. 727–744.

80. Lalli, F., Falchi, M., Romano, G.P., Romolo, A., and Verzicco, R., (2007), Jet-wall interaction in shallow waters, *International Journal of Offshore Polar Engineering*, vol. 17, no. 2, pp. 1–5.

81. Lloyd, P.M., and Stansby, P.K., (1997), Shallow-water flow around model conical islands of small side slope Part II Submerged, *Journal of Hydraulic Engineering*, vol. 123, pp. 1068–1077.

82. Marrocu, M., and Ambrosi, D., (1999), Mesh adaptation strategies for shallow water flow, *International Journal of Numerical Methods in Fluids*, vol. 31, pp. 497–512.

83. Cecchi, M., and Marcuzzi, F., 1999. Adaptivity in space and time for shallow water equations, *International Journal of Numerical Methods in Fluids*, vol. 31, pp. 285–297.

84. Lalli, F., Gallina, B., Mozzi, M., and Romano, G.P., (2004), Interaction between river mouth flow and marine structures: numerical and experimental investigations, in G.H. Jirka, and W.S.J. Uijttewaal (eds), Shallow Flows, Taylor & Francis Group, London.

85. Lalli, F., Bruschi, A., Lama, R., Liberti, L., Mandrone, S., and Pesarino, V., (2010), Coanda effect in coastal flows, *Coastal Engineering*, vol. 57, pp. 278–289.

86. Fischer, H., (1973), Longitudinal dispersion and transverse mixing in open channel flow, *Annual Review of Fluid Mechanics*, vol. 5, pp. 59–78.

87. Booij, R., (1989), Depth-averaged k–ε modeling, in *Proceedings of IAHR XXIII Congress*, pp. 199–206.

88. Lien, S.J., and Ahmed, N.A., (2012), Indoor air quality measurement with the installation of a rooftop turbine ventilator, *Journal of Environment Protection*, vol. 3, no. 11, pp. 1498–1508 (November issue).

89. Lien, S.J., and Ahmed, N.A., (2010), Numerical simulation of rooftop ventilator flow, *Building and Environment*, vol. 45, no. 8, pp. 1808–1815.

90. Wu, C., and Ahmed, N.A., (2011), Numerical study of transient aircraft cabin flow field with unsteady air supply, *AIAA Journal of Aircraft*, vol. 48, no. 6, pp. 1994–2002 (November–December issue).

91. Wu, C., and Ahmed, N.A., (2012), A novel mode of air supply for aircraft cabin ventilation, *Building and Environment*, vol. 56, pp. 47–56.

92. Ahmed, N.A., (2012), Novel developments toward efficient and cost effective wind energy generation and utilization for sustainable environment, *Renewable and Power Quality Journal*, vol. 4, no. 10, pp. 1–23 (April issue).

93. Rashid, D.H., Ahmed, N.A., and Archer, R.D., (2003), Study of aerodynamic forces on rotating wind driven ventilator, *Wind Engineering*, vol. 27, no. 1, pp. 63–72.
94. Shun, S., and Ahmed, N.A., (2008), Utilizing wind and solar energy as power sources for a hybrid building ventilation device, *Renewable Energy*, vol. 33, no. 6, pp. 1392–1397.
95. McDowal, R., (2007), Fundamentals of HVAC systems, SI edition, American Society of Heating, Refrigerating and Air-Conditioning Engineers Incorporated, USA.
96. ASHRAE, (2007), Chapter 56, ASHRAE handbook – HVAC applications, Atlanta, American Society of Heating, Refrigeration, Air-Conditioning Engineers.
97. ASHRAE (2009), Chapter 20, ASHRAE handbook – Fundamentals, Atlanta, American Society of Heating, Refrigeration, Air-Conditioning Engineers.
98. ASHRAE, (2012), Air handling and distribution, in ASHRAE Handbook – HVAC Systems and Equipment, Atlanta, American Society of Heating, Refrigeration, Air-Conditioning Engineers.
99. van Hooff, T., Blocken, B., Defraeye, T., Carmeliet, J., and van Heijs, G.J.F., (2012), PIV measurements and analysis of transitional flow in a reduced-scale model: ventilation by a free plane jet with Coanda effect, *Building and Environment*, vol. 56, pp. 301–313.
100. Hensen, J., Bartak, M., and Frantisek, D., (2002), Modeling and simulation of a double-skin façade system, *ASHRAE Transactions*, vol. 108, no. 2, pp. 1251–1259.
101. Guardo, A., Coussirat, M., Egusquiza, E., Alavedra, P., and Castilla, R., (2009), A CFD approach to evaluate the influence of construction and operation parameters on the performance of active transparent façades in Mediterranean climates, *Energy and Buildings*, vol. 41, pp. 534–542.
102. Baldinelli, G., (2009), Double skin façades for warm climate regions: analysis of a solution with an integrated movable shading system, *Building and Environment*, vol. 44, pp. 1107–1118.
103. Valentín, D., Guardoa, A., Egusquizaa, E., Valeroa, C., and Alavedrab, P., (2013), Use of Coanda nozzles for double glazed façades forced ventilation, *Energy and Buildings*, vol. 62, pp. 605–614.
104. Kim, H.D., Rajesh, G., Setoguchi, T., Matsuo, S., (2006), Optimization study of a Coanda ejector, *Journal of Thermal Science*, vol. 15, pp. 331–336.
105. Guardo, A., Coussirat, M., Egusquiza, E., Alavedra, P., and Castilla, R., (2009), A CFD approach to evaluate the influence of construction and operation parameters on the performance of active transparent façades in Mediterranean climates, *Energy and Buildings*, vol. 41, pp. 534–542.
106. Guardo, A., Coussirat, M., Valero, C., Egusquiza, E., and Alavedra, P., (2011), CFD assessment of the performance of lateral ventilation in double glazed façades in Mediterranean climates, *Energy and Buildings*, vol. 43, pp. 2539–2547.
107. United Nations (2014), United Nations World's population increasingly urban with more than half living in urban areas, New York, prospects-2014.html.
108. By Shannon1 – Own work, CC BY-SA 4.0, https://commons.wikimedia.org/w/index.php?curid=65845951
109. https://commons.wikimedia.org/wiki/File:2009-08-30_09-03_Luang_Prabang_020_Mekong.jpg (This image was originally posted to Flickr by Allie_Caulfield at https://www.flickr.com/photos/28577026@N02/3928271590. It was reviewed on 17 October 2009 by FlickreviewR and was confirmed to be licensed under the terms of the CC-by-2.0)

110. Ellis, W.H., and Gray, D.M., (1966), Interrelationships between the peak instantaneous and average daily discharges of small prairie streams, *Canadian Agricultural Engineering*, pp. 1–39 (February issue).

111. Mekong River Commission (2009), The flow of the Mekong, MRC management information booklet, series, no. 2 (November issue).

112. Venkataraman, P. (1977), Divided flow in channels with bottom openings, *Journal of Hydraulic Engineering*, American Society of Civil Engineers, vol. 103, no. 2, pp. 1900–1904.

113. Nasser, M.S., Venkataraman, P., and Ramamurthy, A.S. (1980), Flow in a channel with a slot in the bed, *Journal of Hydraulic Research, Delft, The Netherlands*, vol. 18, no. 4, pp. 359–367.

114. Ramamurthy, A.S., Zhu, W., and Carballada, B.L. (1994), Flow past a two-dimensional lateral slot, *Journal of Environmental Engineering*, American Society of Civil Engineers, vol. 120, no. 6, pp. 1632–1638.

115. May, D., (2015), Sediment exclusion from water systems using a Coanda effect device, *International Journal of Hydraulic Engineering*, vol. 4, no. 2, pp. 23–30.

116. Wahl, T.L. (2001), Hydraulic performance of Coanda effect screens, *Journal of Hydraulic Engineering*, American Society of Civil Engineers, vol. 127, no. 6, pp. 480–488.

117. Wahl, T.L. (2003), Design guidance for Coanda effect screens, R-03003, U.S. Department of Interior, Bureau of Reclamation.

118. Bureau of Reclamation, U.S. Dept. of the Interior, (1987), *Design of Small Dams*, 3rd edition, pp. 365–371.

119. Kamanbedast, A.A., Masjedi, A., and Assareh, A., (2012), Investigation of hydraulics of flow in bottom intake structures by software modeling, *Journal of Food, Agriculture and Environment*, vol. 10, no. 2, pp. 776–780.

120. Day, T.R., (2006), Coanda effect and circulation control for non-aeronautical applications, Chapter 24, in R.D. Joslin, and G.S. Jones (eds) *Applications of Circulation Control Technologies, Progress in Astronautics and Aeronautics*, vol. 214, pp. 599–613.

121. Englar, R.J., (2006), Pneumatic aerodynamic technology to improve performance and control of automotive vehicles, Chapter 13, in R.D. Joslin, and G.S. Jones (eds) *Applications of Circulation Control Technologies, Progress in Astronautics and Aeronautics*, vol. 214, pp. 357–382.

122. Englar, R.J., (2006), Aerodynamic heat exchanger: a novel approach to radiator design using circulation control, Chapter 14, in R.D. Joslin, and G.S. Jones (eds) *Applications of Circulation Control Technologies, Progress in Astronautics and Aeronautics*, vol. 214, pp. 383–398.

123. Olivotto, C., (2010), Fluidic elements based on Coanda effect, *INCAS Bulletin*, vol. 2, no. 4, pp. 163–172.

124. Allery, C., Guerin, S., Hamdouni, A., and Sakout, A., (2004), Experimental and numerical POD study of the Coanda effect used to reduce self-sustained tones, *Mechanics Research Communications*, vol. 31, pp. 105–120.

5

Coanda Effect in a Human Body

5.1 Cardiovascular Disease

The Coanda effect may be used to explain many of the complex fluid flow phenomena that take place inside a human body. The healthy functioning of a human body is highly dependent on the normal movement of fluid, such as blood or air, within it. The movement of blood is required for the transport of glucose, oxygen, and other nutrients to different parts of the body, while the flow of air through windpipes is essential for breathing or making speech.

Any abnormality in the flow movement within a human body gives rise to various diseases and medical conditions; some of these abnormalities may be linked to the Coanda effect. A greater understanding of this fluidic effect may help medical professionals to develop effective diagnostic procedures and treatments for various ailments and diseases, and in many cases, recommend preventative measures before such diseases take hold.

An important area where the Coanda effect is known to cause problems is in the development of cardiovascular disease. Cardiovascular disease is a major cause of death globally. In the last quarter of a century, global deaths reported from cardiovascular disease have increased significantly, from over 12 million to nearly 18 million [1]. A snapshot of the distribution of deaths around the world [2] is presented in Figure 5.1. An interesting point evident from this figure is that the disease is more prevalent in rich countries than in the poorer parts of the world, a possible after effect of a stressed lifestyle and affluent unhealthy food habits.

Another area where the Coanda effect occurs is during phonation, a process whereby speech or sound is generated. Phonation plays an important role in human communication, and abnormalities associated with its functioning constitute a significant area of medical research.

Today, there is great urgency to find solutions to the cardiovascular- and phonation-related problems, and substantial resources are being invested toward that end. A key strategy of these efforts has been to determine the linkages that exist between the fluid mechanics and the adverse medical conditions. The study of bio-fluid dynamical aspects of the human cardiovascular

	318-925
	926-1,148
	1,149-1,294
	1,295-1,449
	1,450-1,802
	1,803-2,098
	2,099-2,624
	2,625-3,203
	3,204-5,271
	5,272-10233

FIGURE 5.1
A snapshot of the distribution of deaths around the world due to cardiovascular disease based on statistics from WHO. (After: [2].)

and phonation systems have, therefore, become an active area of research, and rapidly growing in importance.

In this chapter we will assess how and where the Coanda effect may show signs of developing in a human body and review some of the works that have been conducted in relation to the cardiovascular and respiratory systems.

5.2 Flow Networks in a Human Body

In the study of the Coanda effect in a human body, it is important to consider the nature of the flow movements of blood and air and the associated flow networks that facilitate such movement.

Since human blood is a very complex fluid, a common approach in fluid mechanical investigation has been to treat blood as an incompressible Newtonian fluid. But because white and red cells remain suspended in it, the blood behaves more like a non-Newtonian fluid at low shear rates during its passage through its vessels. The narrowing of the blood vessels is often the consequence of the accumulation of plaques or fatty materials inside the artery. Such accumulations generally cause a reduction in blood flow resulting in serious circulatory disorders, and in many cases, heart attacks. In other instances, pressure may build up within blood vessels and the vessels may enlarge in diameter at some locations, which may significantly reduce their strength and make them prone to rupture and hemorrhage.

Most of these abnormal changes to the blood vessels appear to occur at the curvatures, junctions, and branches of large and medium arteries.

The characteristics of blood flow through the arteries and their branches, therefore, may hold clues for much of the causes of arterial diseases such as atherosclerosis. But the complexities associated have made the task an extremely difficult one. The studies for both normal and defective vessels have been limited to flow fields of idealized cases of arteries and their branching, and on aspects such as the effects of constricted flow, the wall motion on wall shear stress or the effect of bifurcated angles, and so on [3–5].

Regarding phonation phenomenon, investigation do not become simpler either. The movement of air in a human body during respiration is subjected to significant expansion and compression, as well as vibratory and acoustic effects. These cause the subject matter of phonation difficult to handle experimentally and computationally. Thus, as with cardiovascular studies, idealization and heavy simplifications of the processes involved have become the unavoidable practice.

5.2.1 Analogy of a Human Flow Network with an Engineering Flow Network

A common feature of the pipes or blood vessels is that they tend to branch out or connect with each other, forming very complex networks. To illustrate these networks, three simplified sketches are provided in Figure 5.2 (a)–(c). The first sketch, Figure 5.2 (a), is the simplified sketch

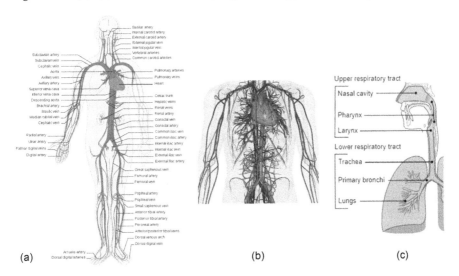

(a) (b) (c)

FIGURE 5.2
Flow network of a human body: (a) simplified sketch of the circulatory system of a human body in anterior view (after: LadyofHats [6]); (b) heart, major veins, and arteries constructed from body scans (after: Public Domain [7]); (c) respiratory system, from the nose to the location of trachea and bronchi. (After: [8].)

of the circulatory system in a human body in anterior view [6]. The second sketch shows the heart along with the arteries and veins constructed from major scans [7]. The third sketch, Figure 5.2 (c), shows the simplified sketch of the human respiratory system [8] that is responsible for breathing and voice production.

Broadly speaking, the windpipe or blood vessel networks can be compared with the flow networks of many engineering devices, such as heat exchangers [9], chemical reactors [10], air-conditioners [11], electronic device coolers [12], fuel cells [13], solar thermal collectors [14], land irrigation systems [15], among others. The design objectives of any engineering device include optimum cost, high performance efficiency, reliability, and durability. To meet these objectives in the engineering networks, uniform flow through them is essential.

Several studies have shown that for a healthy and disease-free existence, a similar requirement of flow uniformity within human flow networks is also called for. Lee et al. [16] have demonstrated that the uniform flow distribution in human bodies reduces pressure losses and flow resistance in vascular or circulatory systems. Zarins et al. [17] have found some correlations to exist between atherosclerosis locations and flow patterns in carotid bifurcations. They have further shown intimal thickening and atherosclerosis to be linked to low wall shear stresses, flow separation, and deviation of flow from the axis of the blood vessels. These results are also supported in the works of Ku et al. [18]; while working on the effect of pulsatile flow on carotid bifurcations, they obtained correlations between the intimal thickness and wall shear stress, and between intimal thickness and flow oscillation.

Several solutions have been proposed to find optimum conditions and configurations that would ensure flow uniformity through a flow network. Of these, the design of flow channel networks based on symmetric bifurcation that can be further extended to generate a larger number of distributed flow streams in a cascading manner [19] appears quite promising. The concept of cascading channel bifurcation has a major advantage in engineering applications since such structures give engineers greater freedom to design optimal configurations at each step and produce uniform flow to meet specific functional requirements. We will, therefore, look in some detail the bifurcation aspect a flow network.

5.2.2 Formation of Channel Flow Networks through Bifurcation

Bifurcation is the first step in the formation of any flow network. This topic has received much attention with regards to hydrodynamic problems [20]. Morphological investigations suggest that the number of branches at each bifurcation point in a vessel tree to be nearly two [21]. Hence a binary bifurcation vessel tree including a main vessel and two branch vessels has been

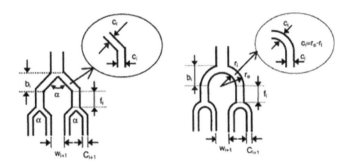

FIGURE 5.3
(a) and (b): Basic flow channel bifurcation structures [19]): (a) tree-bifurcation; (b) circular-bifurcation.

used as the standard model. Two basic flow channel bifurcation structures based on the concept of cascading, namely the tree-bifurcation or circular bifurcation [19], are shown in Figure 5.3 (a) and (b), respectively.

The aforementioned cascading concept of a bifurcated flow network that has been successfully applied to engineering applications may also be used to produce simplified structures during human coronary artery studies. The drawback, however, lies in faithfully representing the real case. Thus, the experimental studies on human flow network bifurcations have been very limited. Apart from the cost and manufacturing constraints, the production of realistic physical test models has never proved easy. The problems are further compounded by the spatial and temporal limitations of the associated equipment [22] that are essential for the acquisition of accurate and reliable data.

Computational fluid dynamics has, therefore, become the main tool of investigation [23]. These studies, too, are hampered by the lack of suitable numerical schemes that need to be validated with reliable data that must come from physical experimentation or real situations. As a result, most of these studies have been based on two-dimensional, idealized [24] or semi-idealized [25] models of coronary arteries and applied to the primary flows and their effects on wall shear stress [26], but very little on the secondary flows [27]. In some studies, the effect of the bifurcation of arteries, which has a significant impact on blood flow characteristics, have been neglected altogether [28], thereby diminishing the prospect of capturing the Coanda effect.

Some attempts have also been made to study bifurcated flows using three-dimensional models. In these computational studies, simplified geometries have been assumed to represent the three-dimensional models as shown in Figure 5.4.

Doutel et al. [29] have used such a model, as shown in Figure 5.4, in their studies of flows in a coronary bifurcation. They approximated the

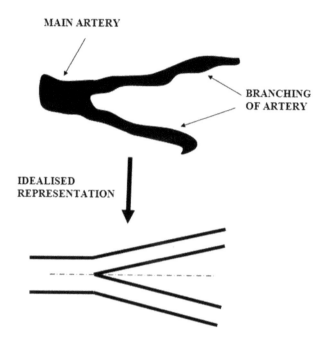

FIGURE 5.4
An artery branching into two is represented by an idealized "Y" shape.

three-dimensional model of the left main trunk of the left coronary artery that bifurcated into the left anterior descending and circumplex arteries that took the appearance of a "Y" shape.

The Y-configuration has also been used in other studies. Yang et al. [30] have used this configuration, as shown in Figure 5.5, to predict the effect of the bifurcation angle on the hemodynamics in the human microvascular system. They found an expanding low-velocity region at the bifurcating region, and velocity augmentation at straight vessels with increasing bifurcating angles. More importantly, they found the flow resistance induced by micro-vessel bifurcation to remain unchanged over their single bifurcation model below a specific diameter (60 μm) regardless of geometric parameters, including the bifurcation angle.

Lamberti et al. [31] have also used the Y-configuration as shown in the schematic of Figure 5.6. They investigated experimentally the adhesion patterns using synthetic microvascular networks. They found that the adhesion pattern propensity increased with larger bifurcating angles with higher adhesion at junctions compared to the straight sections. However, they were unable to observe some expected differences between the upstream and downstream flows in their synthetic channel network flow images, as given in Figure 5.7, which they attributed to their inability to reproduce completely the in vivo blood rheology in their experiment.

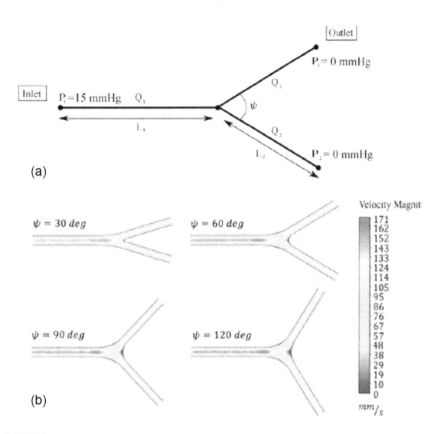

(a)

(b)

FIGURE 5.5
(a) and (b): Effect of the bifurcation angle: (a) 2-D single bifurcation model with no bifurcation angle effect; (b) 3-D single bifurcation models with varying bifurcation angles. (After: [30].)

5.3 Development of the Coanda Effect in a Bifurcating Flow

Chesnutt and Marshall [32] have demonstrated that the distribution of blood cells results in an uneven flow in the downstream channels of a bifurcating flow. From a fluid mechanics point of view, such uneven flow or non-uniformity is something that may give rise to the Coanda flow phenomenon. Thus, concentrating on bifurcation and using the Y-configuration for the flow networks within human body, we can proceed to make some qualitative assessments of where and how the Coanda effect may develop.

For ease of discussion, we have drawn Figure 5.8, which is an idealized bifurcating representation of a flow network taking on a "Y" shape, and labelled its various geometric lengths, diameters, and angles. In the parent tube, we have also introduced a diverging-converging section prior to the

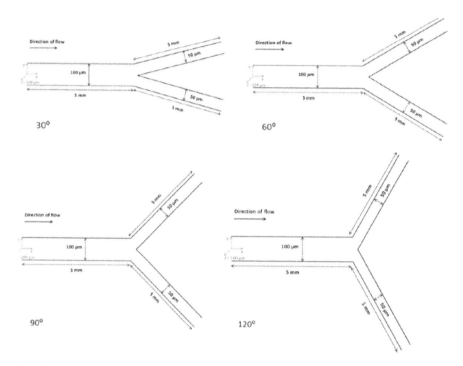

FIGURE 5.6
Schematic of bifurcating synthetic microvascular networks at various bifurcating angles.
(After: [31].)

bifurcation which, if present, can be used to represent the narrowing of the blood vessel from injury, disease, or deposition of plaque originating from fat or cholesterol.

5.3.1 Flow Asymmetry in a Constant Parent Tube Diameter (No Narrowing)

We start our discussion by considering a flow through the simplest configuration, where the parent tube is of a fixed diameter and suffers no narrowing in shape before bifurcation into two symmetrical tubes of equal diameter and angle, as shown in Figure 5.9.

In Figure 5.9 (a), the jet of the fluid flowing from the parent tube is uniform before moving into the two symmetrically bifurcating tubes. We will assume the tubes to be rigid and the jet flow to be subsonic, inviscid, and incompressible.

Under these idealized conditions, the flow in the parent tube will remain uniform and enter the bifurcating tubes with equal and uniform velocity.

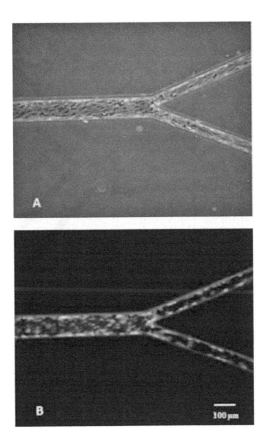

FIGURE 5.7
Images show endothelial cells cultured in synthetic microvascular networks under shear flow conditions. A and are B endothelial cells stained with different colors. (After: [31].)

However, if there is a slight disturbance that causes non-uniformity in the flow of the parent tube, then the symmetry of flow propagation within the parent tube will be broken. The jet will then move more toward one wall than the other, causing a reduction in the upstream area that is feeding the flow. An unequal flow of different velocities will then enter the two bifurcating tubes, one having a higher velocity compared to other. The tube with the lower velocity may not follow the contour of its wall properly, and may even separate from it.

Such flow asymmetry in the parent tube can often be considered as the precursor to the Coanda effect, taking hold in one of the bifurcating tubes in which the flow is moving with a higher velocity. After an initial instability, the flow will become stable and follow the contour of this tube while adhering to its wall, as depicted in Figure 5.9 (b).

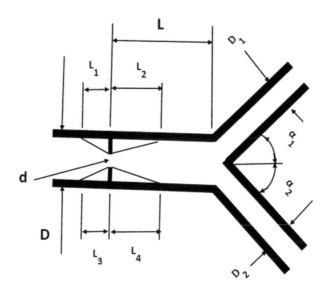

FIGURE 5.8

Flow bifurcation of a parent tube into a "Y" shape. Where: D is the diameter of a parent tube; d is the throat at the divergent–convergent section; L is the length of the parent tube; L_1 and L_3 are lengths of the divergent–convergent section upstream of the throat; L_2 and L_4 are lengths of the divergent–convergent section downstream of the throat; D_1 and D_2 are the diameters of the bifurcated tubes; α_1 and α_2 are the angles of the bifurcated tubes from the parent tube. $(\alpha_1 + \alpha_2)$ is the angle of bifurcation; we will denote this by α such that, $0° < \alpha < 180°$.

5.3.2 Flow Asymmetry in a Non-Constant Diameter Parent Tube

Let us now consider a general flow through a parent tube that has a converging and diverging section prior to bifurcation.

With the assumption of rigid tubes and subsonic, incompressible, and inviscid flow, we can apply the conservation laws of mass and momentum between the flow in the parent tube upstream of the converging section and the throat. We will find an increase in velocity and a decrease in pressure as the flow converges to the minimum area at the throat. A decrease in velocity and an increase in pressure can be expected in the diverging section as the flow moves from the throat to its downstream.

Generally, any abrupt change in the direction of a flow is undesirable. In a subsonic flow, the converging process suffers lower losses than a deceleration process. This is because during deceleration, as the flow expands through a diverging section, it is prone to flow separation, reverse flow, or wake formation, all of which are the sources of losses. To minimize losses in expansion, the lengths of L_2 and L_4 of the divergent section have to be longer compared to the converging lengths of L_1 and L_3. This suggests a preference for a smaller diverging angle for low loss flows. If these lengths of both the

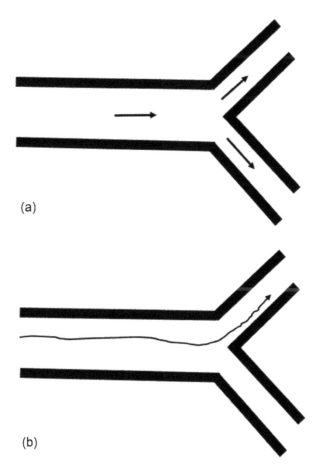

FIGURE 5.9
(a) and (b): Flow bifurcation into a "Y" shape without any constriction in the parent tube: (a) uniform flow; (b) non-uniform flow showing the Coanda effect.

converging and diverging sections are of such scale that the jet flow moves smoothly in the parent tube without suffering losses, then no Coanda effect will be produced during bifurcation.

Let us also look at an extreme case when these four lengths are so small that they can be ignored. Then the converging–diverging section takes the shape of an orifice plate and forms a constriction as shown in Figure 5.10. We will also assume a symmetric bifurcation, implying that the diameter, d, of the constriction is centrally located in the parent tube and D_1 and D_2 are of equal magnitude and their angles from the parent tube, α_1 and α_2, are also equal.

Referring now to Figure 5.10 (a) and (b), when the jet enters from a smaller area of the constriction to a larger area downstream, it will entrain fluid

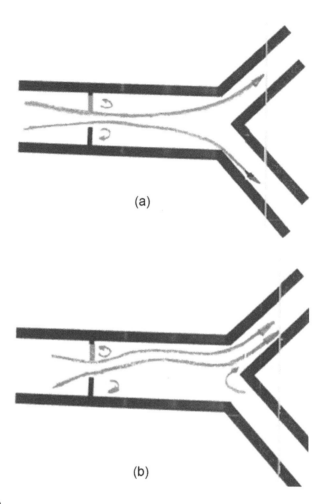

FIGURE 5.10
(a) and (b): Uniform flow with a constriction in the parent tube bifurcating into a "Y" shape: (a) uniform flow; (b) non-uniform flow showing the Coanda effect.

from its surroundings. The sudden removal of fluid in this way will lower the pressure between the jet and the side walls of the parent tube and create reverse flow and the formation of two dead or re-circulating flow zones downstream of the constriction. The formation of the recirculation zones is shown in Figure 5.10 (a). Despite losing some momentum because of the formation of these recirculating zones, the flow is still expected to branch out equally into the bifurcating tubes because of geometrical symmetry, if "L" is of sufficient length. But if there is a slight disturbance in the flow of the parent tube with constriction, the flow will become asymmetric as was the case for the flow without constriction. This scenario, as shown in Figure 5.10 (b), then has the potential to give rise to the Coanda effect.

5.3.3 Flow Asymmetry in the Parent Tube with a Hump

The Coanda effect may also be produced if there is a hump located on any one side of the parent tube. This will have the effect of immediately diverting the flow away from the wall where the lump is located; the symmetry of the flow will be lost. The effect will be more pronounced if the lump is located near the point of bifurcation, as shown in Figure 5.11.

5.3.4 Effect of L, D, and α on Flow Asymmetry

We will now consider the effect of L, the length that extends from the constriction to the bifurcation location, and the constriction diameter, D, for a given velocity of a jet. When L is small, there will be very little Coanda effect. However, as L increases, with little disturbance, the flow may start to become more and more bi-stable even at very low pressure and Reynolds number. The Coanda effect may become more pronounced until reaching a limiting value of L, after which the effect would start to diminish gradually.

Compared to D, L should be substantially larger. This is to ensure that the flow has become fully developed before it bifurcates. As a rule of thumb, L is often taken to be greater than 10 times the value of D (L > 10D).

The diameters, D_1 and D_2, of the bifurcated tubes may also play some role. For the Coanda effect to occur, the diameters of the jet at entry to the bifurcated tubes should be less than bifurcated tube diameters.

Finally, the effect of the bifurcating angles, α_1 and α_2, is considered. The combined value of α_1 and α_2 can range from 0 to 180° (0° < α <180°) where

FIGURE 5.11
Non-uniform flow generated due to a hump in the parent tube and bifurcating into a "Y," shape showing Coanda effect.

$\alpha = (\alpha_1 + \alpha_2)$. Let us assume $\alpha_1 = \alpha_2$, and take the extreme case of α being close to 180°, then the flow from the parent tube will have to turn a corner of nearly 90° to enter the bifurcated tubes. This will be difficult to accomplish without suffering significant pressure and momentum losses. The flow velocity will be reduced, and the pressure will rise, which in turn may produce reverse flow. Under such circumstances, the Coanda effect may be produced irrespective of whether a hump or a constriction is present on the parent tube or not.

We have so far conducted our discussion by concentrating on flow asymmetry in a symmetric configuration where asymmetry may be brought about by some disturbance. We have argued that, if the parent tube flow, with or without constriction, departs from its normal flow trajectory, then the chances for the Coanda effect to occur become high. We have not elaborated on what happens to the flow in the bifurcating tubes, but we expect the process to be similar to the parent flow in subsequent bifurcations. The assumptions of blood vessels or windpipes being rigid and the flow being inviscid are not entirely valid. Our descriptions are qualitative in nature and an oversimplification of a process that is much more complex.

5.4 Coanda Effect in Operative Procedures

Based on the materials presented so far, we can now make some qualitative judgements as to how to avoid the Coanda effect in some of the operative procedures. We will consider some simple steps that can be adopted and avoid the detrimental aspects of Coanda effect with a greater awareness of the flow phenomenon.

We should note, however, that in the previous section, we did not consider how the disturbance in the flow in a parent tube that gives rise to asymmetry may originate. The disturbance may be caused by geometric, physiological flow characteristics, or a combination of both, but also through other factors such as medical conditions or diseases in a human body. In other cases, for example during operative procedures, the Coanda effect may be introduced unwittingly.

In the following sections, we will briefly consider two operative procedures where the Coanda effect, which was not present originally, can manifest: in an aneurysm and in endotracheal intubation.

5.4.1 Aneurysm

We will now consider a scenario when the Coanda effect may occur and produce adverse consequences [33], in the operative treatment of aneurysm. According to the American Heart Association, an aneurysm is said to occur when part of an artery weakens, allowing it to widen abnormally or balloon out.

The causes of aneurysms are not often known. A person may be born with aneurysm, that is, the condition may be genetic, or the condition may arise from smoking, blood pressure, trauma, injury, aortic disease or the aging of the body.

Aneurysms can occur at various places in the human body, the two most common being the aortic aneurysm that occurs in the major artery of the heart, and the cerebral aneurysm that occurs in the brain. Two images of an aneurysm, one in the brain (cerebral aneurysm) [34] and the other in the abdomen [35] are shown in Figures 5.12 (a) and (b), respectively, for illustrative purposes.

Various diagnostic procedures are available to detect aneurysms. Prominent amongst them are non-evasive scans or imaging techniques, such as the computed tomography scan and magnetic resonance imaging technique. The development and role of such techniques in the diagnosis, follow-up, and surgical planning of aortic aneurysms and acute aortic syndromes are well documented and easily available in the available literature [36].

When it comes to the treatment of aneurysms, endovascular therapy or endovascular aortic repair is generally preferred over microsurgical clipping as they are considered to be minimally invasive and cost–time effective. The procedures have been used for the last few decades and are carried out inside blood vessels; they are generally used to treat peripheral arterial disease commonly found in the leg, aorta, or carotid. It is worth noting that not all aortic aneurysms can be repaired with the endovascular aortic repair procedure because the exact location or size of the aneurysm may hamper the safe and reliable placement of the stent graft inside the aneurysm. This would mean resorting to other surgical procedures.

In endovascular aortic repair, a graft is inserted into the aorta to strengthen the aorta without removing the aneurysm. The surgeon first inserts a catheter into an artery in the groin (in the upper thigh) and using the X-ray display, threads the graft, often called a stent graft, into the aorta to the aneurysm. The graft is then expanded inside the aorta, which helps to lock it in place and form a stable channel for blood flow. The graft strengthens the weakened section of the aorta and prevents the aneurysm from rupturing. The endovascular aortic repair process [37] is shown in Figure 5.13.

During the endovascular aortic repair process, care has to be taken to ensure that the graft blends well with its surroundings to ensure smooth and uniform flow through it before it branches out into other smaller arteries. This will help avoid creating any Coanda effect in the arterial branches.

In other operative procedures in the treatment of aneurysms, the surgeon may attempt to isolate the aneurysm from circulating blood. But he has to clip the aneurysm without deforming, distorting, or damaging the adjoining blood vessels and their branches. This makes the procedures extremely delicate and difficult to perform. During this procedure, the Coanda effect, not present originally, may be created unwittingly.

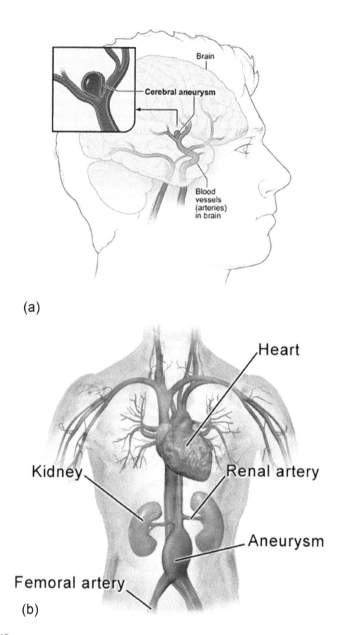

(a)

(b)

FIGURE 5.12
(a) and (b): Images of aneurysms: (a) a cerebral aneurysm (after: [34]) and abdominal aneurysm. (After: [35].)

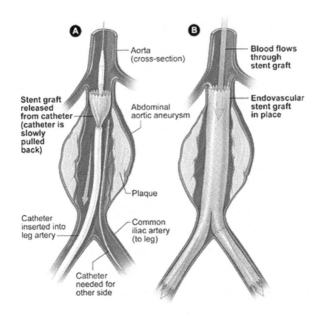

FIGURE 5.13
The placement of an endovascular stent graft in a thoracic aortic aneurysm. (After: US Gov [37].)

To better understand how the Coanda effect may be avoided during clipping, we may conduct our discussion with reference to the cerebral vascular network, shown in Figure 5.2 (b), that supplies blood to the brain. Blood to the anterior (front) and the posterior (back) areas of the brain are supplied by the internal carotid and vertebral arteries, respectively. The right and left vertebral arteries pass through the skull and join to form a single basilar artery. The basilar artery and the internal carotid arteries meet in the form of a ring at the base of the brain. This ring is called the Circle of Willis [38], and is shown in Figure 5.14.

The majority of the cerebral aneurysms are found to be in the anterior segment of the Circle of Willis, and can be found in three main arteries: the internal carotid artery, the middle cerebral artery, and the anterior cerebral artery. There are others in the basilar artery, vertebral basilar, posterior communicating artery, and cavernous carotid artery.

From the Circle of Willis (refer to Figure 5.14), we can easily observe that the internal carotid artery bifurcates into the middle cerebral artery and the anterior cerebral artery, forming the "Y" shape. The bifurcating angle is also large and close to 180°.

It should be noted that aneurysms arising from the internal carotid artery trunk have a higher risk of intraoperative rupture with a large defect in

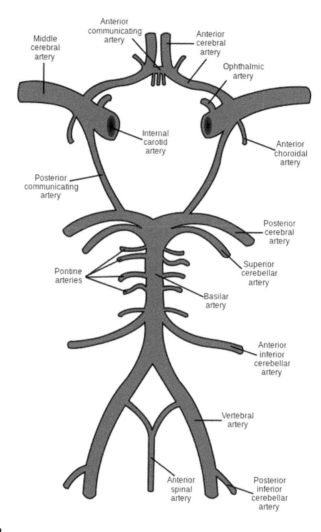

FIGURE 5.14
The anterior and posterior circulations meet at the Circle of Willis, pictured here, which rests at the top of the brainstem. (After: [38].)

the wall due to their very fragile walls and poorly defined necks [39, 40] Therefore, these aneurysms require special attention.

With the advent of microvascular techniques, most parts of the damaged artery wall during aneurysm surgery can be repaired by primary closure or by the bypass technique. However, these methods are not always successful [41].

The Sundt encircling clip has, therefore, been developed to repair defects of some of the blood vessels. While installing this clip, great care would be required to ensure that no constriction is formed in the blood vessel. This will help prevent the formation of the Coanda effect in the vessel. Often this fluidic effect may not be noticed during operative procedures, but it may become evident during the post-operative phase. It is also important to take into consideration the effect of the length, "L," of the internal carotid artery during clipping. Clipping the lower aneurysm will be less risky than the upper aneurysm because of a larger L value. Similar to the case of a hump, noted earlier (refer to Figure 5.11), any aneurysm particularly close to the bifurcation location will have the possibility of reducing the blood flow in the middle cerebral artery, which will divert an accelerated blood flow into the anterior cerebral artery. These outcomes are clearly hazardous.

If a normal clip is used at the junction where the middle cerebral artery and the anterior cerebral artery meet, the curvature will not be affected and no distortion to the configuration will take place, suggesting no formation of the Coanda effect.

Surgery on middle cerebral artery aneurysms can sometimes be quite difficult and also dangerous because of the arteries' wide base, large size, varying thickness and irregular dome characteristics. The majority of these aneurysms are located at the bifurcation or trifurcation of the middle cerebral artery. Pre-bifurcation aneurysms are relatively rare. The knowledge of the anatomical and physiological fact, therefore, is very important when deciding how to treat an aneurysm of the middle cerebral artery [42–44].

When an aneurysm occurs at the trifurcation of the middle cerebral artery, the flow tends to have a symmetrical flow. This diminishes the prospect of the Coanda effect occurring when the clip is applied at right angles to the parent vessel. If, however, the trifurcation is not symmetrical and has primary bifurcation preceded by a secondary bifurcation, then clipping at 90° to the parent vessel is unlikely to yield satisfactory result.

Finally, we look at the clipping of an aneurysm on the anterior cerebral artery, which is one of a pair of arteries on the brain that supplies oxygenated blood to most midline portions of the frontal lobes and superior medial parietal lobes. Here the clipping has to be performed at right angles to the vessel. If the clipping occurs parallel to the vessel, the chances of fluid being entrained from the pericallosal artery are considerable. Another possibility is that a jet may be formed that flows into the contralateral communicating artery. Both outcomes are dangerous may have dire consequences.

5.4.2 Endotracheal Intubation

Endotracheal intubation is another operative procedure during which the Coanda effect may be introduced unintentionally. Endotracheal intubation is an artificial form of ventilation administered to a patient.

To understand the process, let us consider the human respiratory system, as depicted earlier in Figure 5.2 (c). Here the trachea is the windpipe or airway that connects the nose and mouth to the lungs. When a person inhales, it is through the windpipe that air flows into the lungs. Thus, a properly functioning and damage-free trachea is essential for respiration and survival.

Endotracheal intubation is generally performed when a patient is placed on a ventilator during anesthesia or serious illness. In this procedure, a tube called an endotracheal tube, is inserted through the mouth and into the airway of the patient. The endotracheal tube then acts as an artificial windpipe that connects the pharynx. The pharynx branches into the two bronchi to finally connect the lungs to supply air. The intubation process [45] is shown in Figure 5.15.

The flow of air during the artificial ventilation can be compared to the air movement through a tube that bifurcates, as discussed earlier. It is during the bifurcation of air flow that the Coanda effect may occur in the bronchi. If that happens, unequal ventilation to the two sides of the lungs may result.

Qudaisat [46] has reported of an incident of unequal ventilation of the lungs during artificial ventilation administered using an endotracheal tube to a patient undergoing a total knee replacement under general anesthesia.

During each breath of the patient, unusual right and left chest expansions were repeatedly observed. The asynchrony or mismatch was further confirmed when a stethoscope was used to listen to the breathing. The breathing sound in the left lung was heard late in the inspiration whereas the sound in the right lung was heard throughout inspiration. Attempts to adjust the depth of the endotracheal tube did not improve the situation.

So, additional chest X-ray tests were carried out. The tests revealed something interesting. There was a noticeable endotracheal tube deviation with a sudden rightward angulation of 90°. It was speculated that this eccentric dislocation of the endotracheal tube may have initiated the Coanda effect on the jet of gas coming from the ventilator into the endotracheal tube and not exiting symmetrically from its end. Consequently, the jet would have

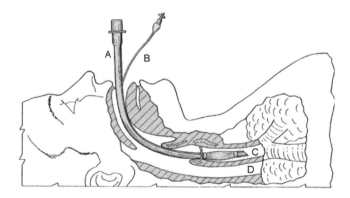

FIGURE 5.15
Endotracheal intubation: diagram of an endotracheal tube that has been inserted into the trachea: A- endotracheal tube; B- cuff inflation tube with pilot balloon; C- trachea; D- esophagus. (After: PhilippN [45].)

adhered to the tracheal wall and moved into the right bronchus, following the tracheal wall contour, generating unequal ventilation.

By carefully deflating the cuff of the endotracheal tube and rotating it by 90° to the left without altering the depth of the tube, the eccentricity introduced during the incubation process was reversed. This appeared to eliminate the Coanda effect. Subsequent tests after this rotational adjustment, synchrony, and equal ventilation in both lungs were found to have been restored, and normal breathing and functioning of the lungs of the patient returned.

5.5 Coanda Effect in Mitral Valve Malfunction and Mitral Regurgitation

Over the years, a large number of numerical and experimental studies have been conducted on blood flowing through various parts of the body. Although there has been several reporting in the literature of the Coanda effect in cardiovascular and echocardiographic assessments, the information they contain are not sufficient to provide in-depth understanding as they lack proper linkage of the fluid mechanical phenomenon with the medical conditions. This underscores the necessity for proper experimental and numerical studies to fill the void.

In this section, we will look at two specific areas where the Coanda effect has been investigated: the first relates to mitral valve malfunction in the human heart and the second to human phonation.

To understand mitral valve malfunctioning, the anatomy of a normal heart and the location of the mitral valve [47] is shown in Figure 5.16. The mitral valve is an organ that allows flow between two chambers of the heart, from the left atrium to the left ventricle. When part of the mitral valve loosely slips backwards into the left atrium, the condition is called mitral valve prolapse. The most common cause of mitral valve prolapse is the chordal elongation of the valve leaflets which, when it is detected, may also be an indication of severe mitral regurgitation.

Mitral regurgitation affects nearly 150 million people worldwide and occurs in men and women equally [48]. Mitral regurgitation is the leaking of blood flow in the heart, otherwise not present in a healthy body. The condition may occur due to pathology affecting one or more components of the mitral valve or due to annular dilatation and geometrical distortion of the sub-valvular apparatus secondary to the left ventricular, associated with cardiomyopathy or coronary artery disease.

Although the disease has been known since the late 1800s [49], not much progress has been made in identifying the major risk factors [50] or mechanisms that trigger it [51]. Sudden death is not uncommon in patients with mitral valve disease. Severe mitral regurgitation may result in chronic

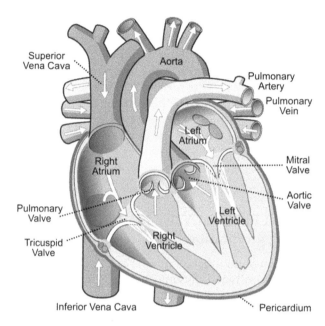

FIGURE 5.16
An anatomy of a normal heart showing the location of mitral valve. (After: [47].)

volume overload and gradual dilatation of the left ventricle, leading to severe heart failure [52–54].

Mitral valve surgery remains the standard of care for patients with symptomatic severe mitral valve regurgitation. Recurrent mitral regurgitation following mitral valve repair is less common, occurring in 10–15% of patients during the first 10 years after surgery [55].

In recent years, a variety of approaches to the percutaneous treatment of primary and secondary mitral regurgitation has been proposed. A task force of the European Society of Cardiology Working Groups on Cardiovascular Surgery and Valvular Heart Disease has been formed to outline steps and recommendations for the treatment and monitoring of mitral regurgitation [56]. Despite this, it remains a fact that there are no medicines currently available to cure this heart valve disease and lifestyle changes, and some medicines can only partially treat or relieve the symptoms and complications.

An echocardiogram is generally used in the diagnosis, management, and follow-up of patients who are suspected to suffer from heart diseases. An imaging technique known as color Doppler echocardiography creates two-dimensional or three-dimensional color images of the intracardiac blood flow patterns, particularly in valvular heart disease, using pulsed-wave or continuous-wave Doppler ultrasound. From such echo Doppler and color flow mapping, the visual images and quantitative information about the size and shape of the heart, as well as the location of the damaged tissues can be determined. In addition, several functional parameters of the heart, such as its pumping capacity, blood flow volume, ejection fraction, and heart relaxation can also be quantified. The information is then used to assess any abnormalities in blood flow between the left and right sides of the heart and the clinical severity of both mitral and aortic regurgitation.

From a fluid mechanics perspective, the malfunctioning of the mitral valve starts with the onset of asymmetries in blood flows. Depending on valve orifice shape and Reynolds number of the flow, this symmetry breaking leads to a bifurcation of the blood flow through the mitral valve or the regurgitant mitral valve, as it is then called. The consequence is a leakage of blood from the left ventricle to the right atrium of the regurgitant mitral valve.

Often, at a low Reynolds number, the leaking of blood flow produces asymmetric expansion of the jet with reduced mass entrainment on the side of the jet close to the surface [57]. Increased shearing forces due to higher spatial velocity differentials in the transverse direction then cause the jet stream to change course and move toward and adhere to the surface, keeping the jet stream from expanding further downstream on the side affected [58, 59]

Since the growth of the jet by entrainment can advance only on the side opposite the surface before moving to intracardiac surfaces such as the ventricular septum, the mitral leaflets, and the left atrial wall adjacent to an eccentric aortic insufficiency, the Coanda effect produced can be viewed on an color echocardiogram [60].

Using echocardiographic color-flow, images of large, central regurgitant jet flow from the left ventricle to the left atrium are generally observed. However, when large, eccentric regurgitant jet flow that adheres to the walls of the left atrium are observed, the effect of the Coanda phenomenon is apparent.

Accurate echocardiographic assessment of Coanda effect occurrence, however, is a significant challenge. The surface-adherent jets of the Coanda effect may be adversely affected by the way a color Doppler scanner is used to compute information from such flows. The jet area in color Doppler flow mapping is also heavily influenced by the instrument setting parameters of frame rate, color gain, pulse frequency, and so forth. Other physiological variables, such as the size of the receiving chamber, diameter of the regurgitant orifice, or driving pressure of regurgitation [61–66], may also affect the quantification of parameters. Different types of jet areas [67, 68] may also be produced when different color Doppler systems are used on the same patient. It has often been found that eccentric regurgitant jets flowing close to the atrial wall or the ventricular septum appear smaller on color Doppler than unbounded jets entering into the center of the ventricle or atrium [69]. This implies that the sizes of the Coanda jets determined in regurgitant flow may be smaller in the diagnostic assessment [70], increasing the risk of gross under-estimation of Coanda jet velocity and regurgitant volume [71, 72]. This is why, in many cases, the severity of the disease is not even detected until at a very advanced stage.

5.5.1 Numerical Simulation of the Coanda Effect in Mitral Regurgitation

Two of the main research foci in mitral regurgitation studies have been to determine the size of the Coanda jets accurately and identify the conditions that trigger the Coanda effect.

Numerical studies using computational fluid dynamics appear to be a better option over physical experimentation when it comes to studying the mitral regurgitation of the heart. The urgency to seek solutions to cardio-vascular diseases, in general, has spurred a significant increase in the study and development of numerical hemodynamic models of the human circulatory system in recent times [73–76]. Other numerical models ranging from lumped models to complex three-dimensional fluid-structure interacting models are also being explored to predict the blood pressure in arteries and the interactions of blood flows with blood pressure walls [77, 78]. Numerical investigations, however, require developing appropriate mathematical models to describe the solid–fluid interactions of the heart and blood flow realistically, and accurate and reliable data for their validation, both of which are currently lacking.

In cardiovascular studies, therefore, an approach that has gained some traction involves finding an analogous situation where some of the salient features of the heart can be reproduced. From this consideration, the blood flow in the heart has been compared to the expansion and contraction of channels flows. This analogy has a major advantage in the sense that a vast amount of literature exists on channel flows. Using the literature and the information contained therein, the prospects of developing useful numerical models for greater understanding of the fluidic phenomenon of the Coanda effect in mitral regurgitation appear brighter. Most of these studies on channel flows, however, have been conducted using incompressible fluid dynamics in planar contraction–expansion settings [79–81], and hence are highly simplified representations of what actually take place.

Since the Coanda effect occurs during the bifurcation of a flow, the instability that occurs during bifurcation of expansion–contraction in the channel flow can provide some useful pointers. Such information can be gathered by seeking hydro-dynamic stability solutions of the incompressible Navier–Stokes equations for a Newtonian and viscous fluid for the channel flows. Numerical methods for hydro-dynamic stability analysis are well established and some are based on linearized eigenvalue problems [82]. Other methods involve direct simulation of the flow and characterization of the flow's asymptotic behaviour [83]. Thus, steady solutions of the Navier–Stokes equations can be obtained for a bifurcating channel flow for different physical and geometric conditions. From these studies, some connection can be drawn between bifurcation theory and linear and non-linear hydrodynamic stabilities.

Chao et al. [84] have suggested that the Coanda effect occurs above a "critical Reynolds number." Drkakais [85] and Revuelta [86] found that this critical Reynolds number was dependent on the expansion ratio, defined as the ratio of the width of the expanding channel to the width of expanded channel, and increased with increasing value of the expansion ratio.

Mishra and Jayaraman [87], while working on asymmetric flows in planar symmetric channels with large expansion ratios, have observed the Coanda effect through the asymmetric solution that produced asymmetric vortices. These vortices remained stable within a certain range of Reynolds numbers, with the asymmetries becoming stronger with an increasing Reynolds number. Wille et al. [88] have attributed the formation of these stable asymmetric vortices in two-dimensional planar expansion to the increase in velocity adjoining one wall that leads to the decrease in pressure near that wall, thereby maintaining the asymmetric nature of the flow.

5.5.2 Hydrodynamic Instability as a Cause of Coanda Effect

Pitton et al. [89] sought to investigate the origin of steady asymmetric flows in a symmetric, sudden-expansion channel flow and use physical experiments to validate them. They started with a consideration of laminar, incompressible, two-dimensional planar contraction–expansion channel flow to investigate the effect of Reynolds number, contraction width, and orifice height on the flow. Thereafter, they went on to examine the influence of three-dimensional geometry on the flow using the same configuration used by Oliveira et al. [90] in their simulation of extensional flows in micro-rheometric devices. The configuration is shown in Figure 5.17

The work of Pitton et al. [89] demonstrated that the flow was characterized by symmetry about the horizontal axis and vertical axis, with the formation of two vortices both upstream and downstream of the contraction. As the Reynolds number increased, the inertial effects began to dominate, the eddies started to diminish in size, and two recirculating regions of equal size began to appear downstream of the expansion point.

The numerical simulations results from Pitton et al [89] are shown in Figure 5.18.

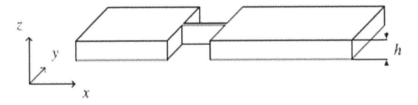

FIGURE 5.17
Schematic representation of the three-dimensional geometry used in a contraction–expansion channel. (After: [89].)

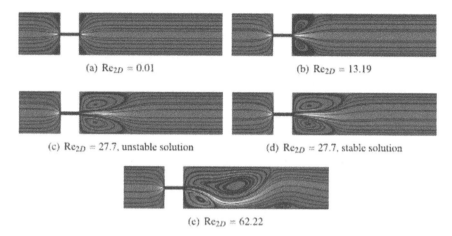

(a) $\mathrm{Re}_{2D} = 0.01$ (b) $\mathrm{Re}_{2D} = 13.19$

(c) $\mathrm{Re}_{2D} = 27.7$, unstable solution   (d) $\mathrm{Re}_{2D} = 27.7$, stable solution

(e) $\mathrm{Re}_{2D} = 62.22$

FIGURE 5.18
Representative snapshots for the 2-D case for the expansion ratio of $\lambda = 15.4$: (a) velocity magnitude and streamlines for $\mathrm{Re}_{2D} = 0.01$; (b) $\mathrm{Re}_{2D} = 13.2$; (c) $\mathrm{Re}_{2D} = 27.7$; unstable solution, (d) $\mathrm{Re}_{2D} = 27.7$; stable solution, (e) $\mathrm{Re}_{2D} = 62.2$. (After: Pitton, Quaini, and Rozza, 2017 [89].)

The gradual transformation of the flow can be observed in Figures 5.18 (a)–(e). Interestingly, beyond the point of bifurcation, two solutions were observed: one is a symmetric but unstable solution, as seen in Figure 5.18 (c); the other is a slightly asymmetric but still a stable solution, as is seen in Figure 5.18 (d). The formation of stable asymmetric vortices in two-dimensional planar expansion is attributed to the Coanda effect. The asymmetric solution remains stable for a certain range of Re_{2D} but with further increases in the Reynolds number, the asymmetries also increase in strength.

From the above results, Pitton et al. [89] concluded that unique symmetric flow, which existed at a small Reynolds number, was not stable at larger Reynolds numbers, i.e., a pitchfork bifurcation formed so that two stable asymmetric steady flows occurred. In fact, in their study, at larger Reynolds numbers, they found eight asymmetric stable steady solutions, with the possibility of another seven unstable solutions. At sufficiently high Reynolds numbers, they also found time-periodic solutions from which they were able to confirm the existence of a Hopf bifurcation. At higher Reynolds numbers, the flow eventually became time-dependent and there was experimental evidence that this was associated with three-dimensional effects.

From the works of Oliveira et al. [90], Cherdron et al. [91], and Chiang et al. [92], it was inferred that the critical Reynolds number for the symmetry-breaking not only varied with the expansion ratio but also with the aspect ratio. At a fixed expansion ratio, but decreasing aspect ratio, the end wall influence became more important and the critical Reynolds number increased. At moderate aspect ratios, the flow was highly three-dimensional but remained steady with time.

Tsai et al. [93] investigated the capabilities and limitations of two-dimensional and three-dimensional methods in modelling the fluid flow in sudden expansion channels. They found complex spiraling structures in three-dimensional flows that were not closed, recirculating cells as was observed in two-dimensional flows. They also found that the flow resembled a two-dimensional flow only for very large aspect ratios.

The theoretical study of Lauga et al. [94] found highly three-dimensional flows at very low aspect ratios. The numerical and experimental studies reported by Oliveira et al. [90] showed strong three-dimensional effects appearing for low aspect ratios that inhibited the wall-hugging or Coanda effect observed in geometries with high aspect ratios at the same Reynolds number. This suggested that the eccentric regurgitant jets, such as the one seen in the mock heart chamber in Figure 5.19 (a), occurred at large expansion and aspect ratios, or in other words, when the regurgitant orifice is narrow and long.

It is interesting to note the claim of Pitton et al. [89] that, following their finding that asymmetry arises at a symmetry-breaking bifurcation, their medical collaborators at the Houston Methodist DeBakey Heart & Vascular Center were also able to reproduce the Coanda effect in a mock heart chamber, shown in Figure 5.19 (b) [95].

5.6 Coanda Effect in Phonation

The Coanda effect may be observed in the human phonatory process and have significant medical implications. Healthy speakers use a wide pitch range during phonation, whereas people with vocal problems often have limited pitch ranges, such as the predominantly high vocal pitch due to laryngeal postural problems during phonation in dysphonic patients [96, 97].

The phonatory process, or voicing, occurs when air is expelled from the lungs through the glottis, creating a pressure drop across the larynx (refer to Figure 5.13). When this drop becomes sufficiently large, the vocal folds start to oscillate. The motion of the vocal folds during oscillation is mostly lateral with practically no motion along the length of the vocal folds [98]. The oscillation of the vocal folds modulates the pressure and flow of the air through the larynx. A picture depicting the Glottis [99] and other components surrounding it is shown in Figure 5.20.

The larynx produces a harmonic series of sound that consists of a fundamental frequency and harmonic overtones that are multiples of the fundamental frequency [100]. The sound thus produced excites the vocal tract and produces the individual speech sounds.

The fundamental frequency can be varied by several means, such as increasing the tension in the vocal folds through contraction of the cricothyroid

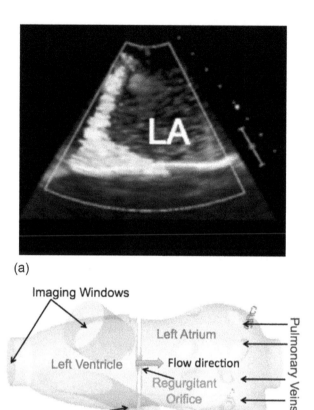

(a)

(b)

FIGURE 5.19

(a) Two-dimensional Doppler echocardiographic image of the regurgitant jet in the mock heart chamber, and (b) geometry of the mock heart chamber with the divider plate between mock left ventricle and mock left atrium (LA). (After: Pitton, Quaini, and Rozza, 2017 [89].)

muscle or through the pressure drop across the larynx. The pressure drop is mostly affected by the pressure in the lungs, and the distance between the vocal folds.

There are the two main theories that can be used to explain the initiation of vibration of the vocal folds. These are the myoelastic theory and the aerodynamic theory [101]. The two theories are not contradictory but, in fact, somewhat complimentary when it comes to explaining this initiation.

According to the myoelastic theory, when the vocal cords are brought together and breath pressure is applied to them, the cords remain closed until the pressure beneath them is sufficient enough to push them apart. This allows air to escape thereby reducing the pressure which makes the

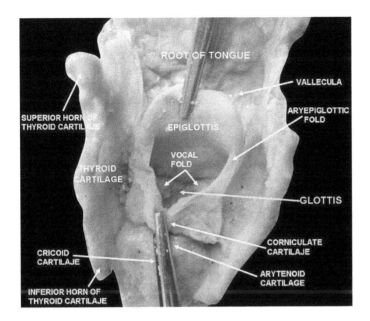

FIGURE 5.20
Image of the glottis. (After: [99].)

muscle tension recoil and pull back the vocal folds. The pressure builds up again, which pushes the cords apart, and the cycle repeats itself.

The aerodynamic theory is similar to the Bernoulli's flow principle applied to a convergent–divergent configuration. According to the Bernoulli's principle, when a stream of breath flows through the glottis, a push occurs during glottal opening, when the glottis is convergent; and a pull occurs during glottal closing, when the glottis is divergent. Due to the push–pull effect on the vocal fold tissues, transfer energy from the airflow to the vocal fold tissues [102], takes place and gives rise to self-sustained oscillation. During glottal closure, the air flow is stopped until breath pressure pushes the folds apart and the flow starts again, and repeats the cycle [103].

There is still a lack of understanding of the geometric features and flow parameters that may have a significant effect on the glottal jet deflection. Some useful observations can, nevertheless, be made from the findings of some of these studies, particularly with regard to the shape, jet and oscillation of the glottis.

5.6.1 Symmetry Breaking in Glottal Flow

The convergent–divergent shapes of the glottis during glottal opening and closing have the potential of producing symmetry-breaking instabilities in the divergent passage [104–106] that can produce the Coanda effect. The characteristics of the glottal jet resembles more a flow that undergoes sudden expansion [107]; and this, as we have noticed earlier when discussing instabilities associated with bifurcation and blood flow network, would make the Coanda jet occurrence dependent on some critical Reynolds number.

The oscillation of the glottal jet caused by its speed variation from nearly zero to a peak velocity that can be as large as 30 m/s over a glottal cycle may also produce unsteadiness and have significant implications for its deflection behavior [108]. Even the presence of small fluctuations in the velocity of the surrounding fluid or the deceleration of the shear layer [108] may be sufficient to encourage or boost asymmetries in the glottal jet.

The inherent unsteady nature of the glottal jet combined with the fact that it interacts with a flow that may already have a certain amount of asymmetry may result in a complex flow where the task of predicting or capturing the attributes of the Coanda effect under simple experimental or numerical settings becomes a difficult proposition.

5.6.2 Numerical Simulation of Phonation

Numerical simulation is now widely used as an efficient tool for clarifying the speech production process[109–116]. A two-mass model developed by Ishizaka and Flanagan [109] and improved models [110, 111] based on this model, because of their simplicity in structure, have become popular in the study of speech production and fluid motion in the larynx.

Nomura and Funada [112] numerically simulated glottal flows on the basis of a two-dimensional rigid wall model of the larynx. They found the glottal flow to be an unsteady asymmetric flow similar to the physically measured flows. They also found that the unsteady flow increased the intensity of the sound in the presence of false vocal folds. Numerical studies by Zhang et al. [113] considered the effect of the false vocal folds on the speech production process using the axisymmetric, forced-vibrating, true vocal fold model. They found free vocal folds generated additional sources in speech waves when impinged upon by the glottal jet.

Most of the above models are, however, not suitable for the description of the complex phenomena encountered [114, 115] because of their limitation with respect to the degree of freedom of fluid motion in the larynx. The models are also unable to provide accurate converging and diverging as well as uniform glottis shapes during any vibration cycle of the true vocal folds. These are apparent from the numerical simulations conducted by Iijima et al. [116] and Liljencrants [117], who concluded that the pressure distributions on the glottal surface were greatly affected by both the shape and minimal diameter of the glottis. Although the authors found asymmetric flows similar to those observed via measurements [118] in their simulation, Iijima et al. [116] were unsure whether their models had captured the asymmetric feature or if there were numerical errors in computation.

A number of other studies have shown that the large-scale asymmetric glottal jet deflection in the medial-lateral direction does occur during phonation both inside the glottis [119] as well as in the supra-glottal region. In many instances, the glottal jet has been found to exhibit stochastic cycle-to-cycle variations in its trajectory [120].

A number of studies suggest asymmetry in the glottal flow has a clear impact on pressure losses, vocal-fold vibration, and sound production [121]. While these studies have implied the Coanda effect to be the cause for glottal jet deflection, other studies have doubted whether it is possible to generate asymmetries that give rise to the Coanda effect in phonatory jets [122].

There are some works where the divergent shape of the glottis is considered to be critical to the asymmetric glottal jet deflection [123], while a different view is expressed in other works where the supra-glottal flow field is thought to play a vital role in driving glottal jet deflection [124].

All these suggest that despite the good volume of work produced on several aspects of phonation, the underlying flow physics remains unclear.

5.6.2.1 Solid–Fluid Interaction Model

Several schemes of solid fluid interaction model have been developed based on the Arbitrary Lagrangian–Eulerian method. These schemes allow the coupling of the Eulerian fluid field with the Lagrangian mechanical field. References to the applications of this method can be found in the works of Hirt et al. [125], Hughes et al. [126], and Wall et al. [127]. An important feature of the Arbitrary Lagrangian–Eulerian method involves keeping the same topology of the fluid grid but allowing slight deformations in accordance with the mechanical displacement along the common fluid–solid interface.

Various attempts have been made to develop the fixed grid or the immersed boundary method to avoid mesh deformation or re-meshing [128–130]. But the real drawback of the immersed boundary method lies in the fact that it is virtually impossible to decouple the physical domain from the fictitious computational domain or avoid the need of additional flow computation in the fictitious domain. Consequently, attempts [131] have been made to combine the benefits of Arbitrary Lagrangian–Eulerian and inverse boundary method toward a partitioned, iterative, coupled scheme based on the extended finite element and distributed Lagrange multiplier or fictitious domain method. These methods, however, require substantial computational resources.

An early continuum mechanical model that has been used in phonation studies can be found in [132], which used the finite element technique to compute vocal fold vibrations. This method consisted of a simple two-dimensional model where the mechanical field was discretized with finite elements and the fluid forces were obtained using Bernoulli's equation. Using a half larynx continuum mechanical model, Thompson et al. [133] have attempted to investigate the causes of self-sustained vocal fold oscillations and found them to originate from the cyclic variation resulting from the convergent to a divergent shape glottis profile. Tao et al. [134] used a self-oscillating finite element model to simulate the impact pressures of vocal folds.

In a subsequent work, Tao et al. [135] constructed a composite two-dimensional model from the Navier–Stokes equations and a two-mass vocal fold description that was used to study the aerodynamics in a vibratory glottis and the vocal fold vibration. Using the model, they were able to predict self-oscillations of the coupled glottal aerodynamics and the vocal cord system. They also found the Coanda effect occurring in the vibratory glottis and concluded that the Coanda effect was responsible for the asymmetric flow in the glottis and the difference in the driving force on the left and the right vocal folds.

One of the first fluid–solid interaction models with fully resolved fluid can be found in [136]. Using this two-dimensional setup and where the inverse boundary method was applied for fluid–solid interaction, the authors were able to simulate significant jet symmetry that attached to either side of the pharynx wall, signifying the occurrence of the Coanda effect.

An attempt to develop a three-dimensional fluid–solid coupled model based on the finite element method with coarse grid can be found in [137]. The authors had some success in demonstrating the self-sustained vocal fold oscillations in their simulations. Overall, full three-dimensional coupled simulations of human phonation are still not well developed.

5.6.2.2 Solid–Acoustic Interaction Model

The aero-acoustic of phonation has been investigated by several authors. Amongst them, Zhao et al.'s works [138, 139] involved the simulation of sound from confined pulsating axisymmetric jets and investigation into the mechanism of sound generation mechanisms while Zhang et al. [140] examined sound generation through glottis-shaped orifices. These studies essentially used a rigid pipe under forced vibration to model the aero-acoustic aspects of the aerodynamic sound produced. In these studies, the fluid–solid interaction was neglected, aiming only on the fluid–acoustic coupling based on Lighthill's acoustic analogy, and solved using the integral method of Ffowcs-Williams and Hawkins [141]. The results from the Ffowcs-Williams and Hawkins-based method were found to agree well with the results obtained using direct numerical simulations that solved the compressible Navier–Stokes equations. In addition to these studies, a theoretical approach was attempted by Crane [142] to develop an acoustic source model. This approach was based on prescribing a jet profile using a train of vortex rings that was then applied to an axisymmetric model of the vocal tract.

Because of the inconsistencies in amplitudes, length, and frequency scales, the monolithic numerical approach has proved hard to implement in acoustics or to find an optimal numerical grid for all scales [143]. This has made the task of obtaining direct solutions for acoustics and flow of the compressible

Navier–Stokes equations, extremely difficult, particularly for low Mach numbers [144].

5.6.2.3 Solid–Fluid–Acoustic Interaction Hybrid Model

Thus, most researchers have either focused separately on the fluid–solid interaction or fluid–acoustic interaction. Human phonation can be viewed as a multi-field phenomenon where the physical fields lie in the continuum of mechanical fields, and where interactions of not only solid and fluid mechanics but also of acoustics take place. Hence a more comprehensive method entailing all the three-fold interactions is desired. There has, unfortunately, been very limited success in developing complete coupled systems that take into account all the fluid–solid–acoustic interactions.

The challenge for such three-field problems lies primarily in the computation of the flow-induced sound. Hybrid methods have, therefore, been developed where the fluid–acoustic interactions can be modelled using Lighthill's acoustic analogy [145, 146] and achieve the solid–acoustic coupling by forcing the continuity of surface velocities in a normal direction. The fluid–solid–acoustic relationship in mathematical modelling and discretization of these hybrid methods is shown in Figure 5.21.

As can be seen from Figure 5.21, the first step of these hybrid methods consists of obtaining a fluid mechanical solution. This is accomplished by conducting flow simulation on a fluid domain including all structures and relevant sound sources. The fluid field is modelled using incompressible, unsteady Navier–Stokes equations, while the solid field uses Navier–Stokes equations modelling linear elasticity and geometric non-linearity. This is then followed by an acoustic solution step where acoustic sources and sound propagation are modelled using an inhomogeneous wave equation based on Lighthill's analogy [145, 146] and extended [147–151] to account for mechanical-acoustic coupling.

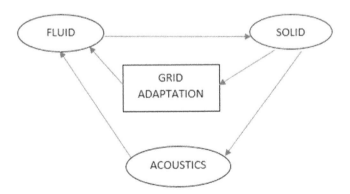

FIGURE 5.21
A simplified fluid–solid–acoustic interaction flow chart.

Zorner et al. [152] and Link et al. [153] are some of the few researchers who have attempted to apply the hybrid concept to larynx flow in human phonation studies. The work of Link et al. [153] also demonstrated the potential of the hybrid approach to capture the Coanda effect in the simulation of larynx flow.

Those authors used the same geometric assembly of the vocal folds that were used by Thompson et al. [133] to represent the vocal folds of the human larynx. The view of the larynx model is shown in Figure 5.22 where the air movement was from the left to the right. The fluid domain was subdivided into two sub-domains: Ω_{ALE} and Ω_{Euler} (see Figure 5.22). The fluid–solid interaction took place in the Ω_{ALE} domain. This part of the mesh had to be adapted within each fluid–solid iteration while the remaining part of the fluid domain Ω_{Euler} required no mesh adaption.

From the fluid–solid–acoustic-coupled results, a qualitative assessment of the development and impact of the Coanda effect were possible. These will be discussed in more detail.

The development of the Coanda effect will be discussed first. This was explored for a fixed glottis width by assuming a constant pressure drop along the glottis. Such a pressure drop may arise within a human body due to the contraction of the lungs. The authors found it advantageous to constrain the volume-induced sound source to a single frequency, which was achieved by prescribing an oscillating pressure drop [154].

The flow boundary conditions were assumed symmetric with symmetric oscillation frequency near the jet front and shear layer. This resulted in symmetric jets forming first. Thereafter, as the shear layer developed, Kelvin–Helmholtz instability set in and jet oscillations started to appear downstream. When the instability moved upstream in time, the corresponding oscillations

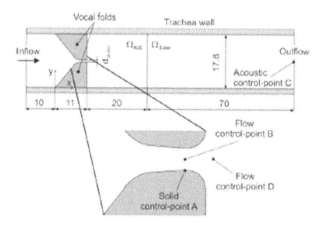

FIGURE 5.22
Computational larynx model. (Distances in mm) (After: [153].)

increased further, and the flow became asymmetric. Finally, the oscillating jets attached to one side of the trachea wall randomly.

Using the velocity magnitude plots around the deformed vocal folds for the first four cycles, Link et al. tracked the development of the Coanda effect. This is shown in Figure 5.23.

The mechanical deformations of the fluid-induced vibrations for each deformation cycle were viewed to start with a forward movement of vocal folds. From an assumed initial state, the vocal folds first deformed and tended to close the gap, making the glottis shape convergent, and then subsequently opened to become divergent. During the back movement of the folds, the shape again became convergent and the cycle repeated continuously as also observed in other experiments [155].

The fact that the flow became asymmetric and randomly attached to one side of the trachea wall may provide a significant contribution to the sound. The region just downstream from the glottis was also found to be a sound source worth noting because strong shear layer regions existed in both.

The velocity field generated large acoustic nodal loads that also represented acoustic point sources and were deemed to be the consequence of highly unsteady flows downstream of a shape-changing glottis.

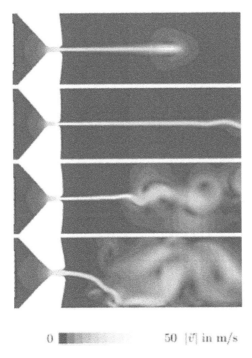

0 ██████ $50 \quad |\vec{v}| \text{ in m/s}$

FIGURE 5.23
Development of the Coanda effect. (After: [153].)

The behavior of flow at control points A, B, C, and D, whose locations can be found in Figure 5.22, when examined in terms of power spectra over a frequency range was used to ascertain the acoustic impact of the Coanda effect.

Under an asymmetric flow condition, i.e., when the Coanda effect was present, the power spectra of the displacement and velocity in the cross flow direction showed significant fluctuations with no single dominating frequency, but in the streamwise direction the existence of a single dominant frequency appeared. The power spectra of the acoustic pressure showed fluctuating signals but also a single dominant frequency under Coanda effect that disappeared in a symmetric flow, replaced by fluctuating signals. The acoustic effects for control point C are shown in Figures 5.24 (a) and (b).

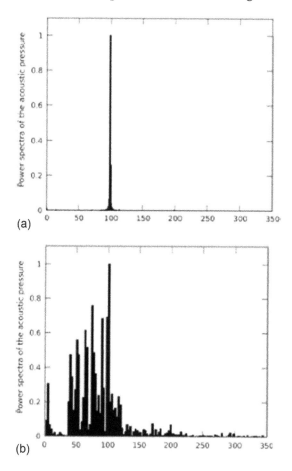

FIGURE 5.24

(a) and (b): Flow behavior at control point C: (a) acoustic pressure at point C under asymmetric condition, i.e., with Coanda effect; (b) acoustic pressure at point C under symmetric condition, i.e., with no Coanda effect. Note: the horizontal axis is frequency (in Hz) (After: [153].)

It was inferred from these studies that the nodal acoustic loads, and therefore, the acoustic field, depended to a great extent on the air jet dynamics for a developed Coanda effect. Similar Coanda effects have also been found in other phonation studies [155].

Although the above studies have shown the Coanda effect to have major impact on the level of sound generated, they have failed to establish how the quality of sound is affected by the fluidic effect.

5.7 Concluding Remarks

A human body is a living entity with an innumerable number of variables that constantly keep changing with time. This makes any investigation into the functions of its various organs and their interactions with each other an extremely difficult undertaking. Thus, despite enormous resources being spent in the health sector worldwide, progress in establishing the desired linkages between the fluid mechanical flow and its medical impacts has been slow. The state-of-the-art of research using physical experimentation and numerical simulation as tools of investigation, particularly in relation to Coanda effect may, therefore, be considered to be at an infant stage.

But there are reasons to be optimistic for the future. We have seen the development of several two-dimensional flow schemes that can simulate the Coanda effect in blood flow networks and the identification of the flow instabilities associated with them. We have also seen the emergence of hybrid solid-fluid-acoustic interaction numerical schemes that have successfully captured the occurrence of Coanda effect in phonation. It is true that they have been achieved under very simplified conditions, but the fact that researchers have been able to capture the main attributes of factors that give rise to the fluidic effect providing the basis for future progress is reason for excitement. It is only a matter of time before realistic, fully three-dimensional, time-dependent numerical solutions and physical experimentation techniques will be developed that will open up new doors for better diagnosis, better control over medical conditions, and overall better health care.

References

1. GBD (2013) Mortality and causes of death, collaborators, (17 December 2014), Global, regional, and national age-sex specific all-cause and cause-specific mortality for 240 causes of death, 1990–2013: a systematic analysis for the Global Burden of Disease Study (2013), *Lancet*, vol. 385, no. 9963, pp. 117–71.

2. http://creativecommons.org/licenses/by-sa/4.0

3. Young, D.F., (1968), Effect of a time-dependent stenosis on flow through a tube, *Journal of Engineering for Industry*, vol. 90, no. 2, pp. 248–254.

4. Chakravarty, S., and Senn, S., (2006), A mathematical model of blood flow and convective diffusion processes in constricted bifurcated arteries, *Korea-Australia Rheological Journal*, vol. 18, no. 2, pp. 51–65.

5. Shaw, S., Gorla, R.S.R., Murthy, P.V.S.N., and Ng, C.O., (2009), Pulsatile casson fluid flow through a stenosed bifurcated artery, *International Journal of Fluid Mechanics Research*, vol. 36, no. 1, pp. 43–63.

6. https://commons.wikimedia.org/wiki/File:Circulatory_System_en.svg (This work has been released into the public domain by its author, LadyofHats)

7. Public Domain, https://commons.wikimedia.org/w/index.php?curid=35786381

8. Lord Akryl, Source: https://www.cancer.gov/; https://upload.wikimedia.org/wikipedia/commons/9/9f/Illu_conducting_passages.svg

9. Bassiouny, M.K., and Martin, H., (1984), Flow distribution and pressure drop in plate heat exchangers, U-type arrangement, *Chemical Engineering Science*, vol. 39, pp. 693–700.

10. Kalisz, S., and Pronobis, M., (2005), Influence of non-uniform flow distribution on overall heat transfer in convective bundle of circulating fluidized bed boiler, *Heat Mass Transfer*, vol. 41, pp. 981–990.

11. Van Gilder, J.W., and Schmidt, R.R., (2005), Airflow uniformity through perforated tiles in a raised-floor data center, in *Proceedings of IPACK2005*, ASME InterPACK, San Francisco, CA, Paper number IPACK2005-73375.

12. Kim, S.Y., Choi, E.S., and Cho, Y.I., (1995), The effect of header shapes on the flow distribution in a manifold for electronic packaging applications, *International Communications in Heat Mass Transfer*, vol. 22, pp. 329–341.

13. Li, X., and Sabir, I., (2005), Review of bipolar plates in PEM fuel cells: flow-field designs, *International Journal of Hydrogen Energy*, vol. 20, pp. 359–371.

14. Weitbrecht, V., Lehmann, D., and Richter, A., (2002), Flow distributors in solar collectors with laminar flow condition, *Solar Energy*, vol. 73, pp. 433–441.

15. Ascough, G.W., and Kiker, G.A., (2002), The effect of irrigation uniformity on irrigation water requirements, *Water South Africa*, vol. 28, pp. 235–241.

16. Lee, J., Lorente, S., Bejan, A., and Kim, M., (2009), Vascular structures with flow uniformity and small resistance, *International Journal of Heat Mass Transfer*, vol. 52, pp. 1761–1768.

17. Zarins, C.K., Giddens, D.P., Bharadvaj, B.K., Sottiurai, V.S., Mabon, R.F., and Glagov, S., (1983), Carotid bifurcation atherosclerosis: quantitative correlation of plaque localization with flow velocity profiles and wall shear stress, *Circulation Research*, vol. 53, pp. 502–514.

18. Ku, D.N., Giddens, D.P., Zarins, C.K., and Glagov, S., (1985), Pulsatile flow and atherosclerosis in the human carotid bifurcation. Positive correlation between plaque location and low oscillating shear stress, *Arteriosclerosis, Thrombosis, and Vascular Biology*, vol. 5, pp. 293–302.

19. Liu, H., Li, P., and Wang, K., (2015), The flow downstream of a flow channel for uniform flow distribution via cascade flow channel bifurcations, *Applied Thermal Engineering*, vol. 81, pp. 114–127.

20. Dijkstra, H., Wubs, F., Cliffe, A., Doedel, E., Dragomirescu, I., Eckhardt, B., Gelfgat, A., Hazel, A., Lucarini, V., Salinger, A., Phipps, E., Sanchez-Umbria, J., Schuttelaars, H., Tuckerman, L., and Thiele, U., (2014), Numerical bifurcation

methods and their application to fluid dynamics: analysis beyond simulation, *Communications in Computational Physics*, vol. 15, no. 1, pp. 1–45.

21. Zamir, M., (1976), Optimality principle in arterial branching, *Journal of Theoretical Biology*, vol. 62, no. 1, pp. 227–251.
22. Buchmann, N.A., Atkinson, C., Jeremy, M.C., and Soria, J., (2011), Tomographic particle image velocimetry investigation of the flow in a modeled human carotid artery bifurcation, *Experiments in Fluids*, vol. 50, pp. 1131–1151.
23. Gijsen, F.J., Schuurbiers, J.C., van de Giessen, A.G., Schaap, M., van der Steen, A.F., and Wentzel, J.J., (2014), 3D reconstruction techniques of human coronary bifurcations for shear stress computation, *Journal of Biomechanics*, vol. 47, pp. 39–43.
24. Chaichana, T., Sun, Z., and Jewkes, J., (2011), Computation of hemodynamics in the left coronary artery with variable angulations, *Journal of Biomechanics*, vol. 44, pp. 1869–1878.
25. Zhang, J.M., Chua, L.P., Ghista, D.N., Zhou, T.M., and Tan, Y.S., (2008), Validation of numerical simulation with PIV measurements for two anastomosis models, *Medical Engineering and Physics*, vol. 30, pp. 226–247.
26. Razavi, M.S., and Shirani, E., (2013), Development of a general method for designing microvascular networks using distribution of wall shear stress, *Journal of Biomechanics*, vol. 46, pp. 2303–2309.
27. Chiastra, C., Morlacchi, S., Gallo, D., Morbiducci, U., Cardenes, R., Larrabide, I., and Migliavacca, F., (2013), Computational fluid dynamic simulations of image-based stented coronary bifurcation models, *Journal of the Royal Society, Interface/ the Royal Society*, vol. 10, no. 84.
28. van der Giessen, A.G., Groen, H.C., Doriot, P.A., de Feyter, P.J., van der Steen, A.F., van de Vosse, F.N., Wentzel, J.J., and Gijsen, F.J., (2011), The influence of boundary conditions on wall shear stress distribution in patients specific coronary trees, *Journal of Biomechanics*, vol. 44, pp. 1089–1095.
29. Doutel, E., Carneiro, J., Campos, J.B.L.M., and Miranda, J.M., (2018), Experimental and numerical methodology to analyze flows in a coronary bifurcation, *European Journal of Mechanics/B Fluids*, vol. 67, pp. 341–356.
30. Yang, J., Pak, Y.E., and Lee, T.R., (2016), Predicting bifurcation angle effect on blood flow in the microvasculature, *Microvascular Research*, vol. 108, pp. 22–28.
31. Lamberti, G., Soroush, F., Smith, A., Kiani, M.F., Prabhakarpandian, B., and Pant, K., (2015), Adhesion patterns in the microvasculature are dependent on bifurcation angle, *Microvascular Research*, vol. 99, pp. 19–25.
32. Chesnutt, J.K.W., and Marshall, J.S., (2009), Effect of particle collisions and aggregation on red blood cell passage through a bifurcation, *Microvascular Research*, vol. 78, pp. 301–313.
33. Robinson, J.L., and Roberts, A., (1972), Operative treatment of aneurysms and Coanda effect: a working hypothesis, *Journal of Neurology, Neurosurgery, and Psychiatry*, vol. 35, pp. 804–809.
34. National Institutes of Health. https://upload.wikimedia.org/wikipedia/commons/8/80/Cerebral_aneurysm_NIH.jpg
35. BruceBlaus, Own Work, CC BY-SA 4.0, https://commons.wikimedia.org/w/index.php?curid
36. Walker, B.R., Colledge, N.R., Ralston, S.H., and Penman, I., (2014), *Davidson's Principles and Practice of Medicine*, 22nd Edition, Churchill Livingstone/Elsevier.

37. US Gov, Public Domain, https://commons.wikimedia.org/w/index.php?cur id=12837854

38. Rhcastilhos – Gray519.png, Public Domain, a. https://commons.wikimedia.org/ w/index.php?curid=1597012

39. Abe, M., Tabuchi, K., Yokoyama, H., and Uchino, A., (1998), Blood blister like aneurysm of the internal carotid artery, *Journal of Neurosurgery*, vol. 89, pp. 19–424.

40. Nakagawa, F., Kobayashi, S., Takemae, T., and Sugita, K., (1996), Aneurysm protruding from the dorsal wall of the internal carotid artery, *Journal of Neurosurgery*, vol. 65, pp. 303–308.

41. Sanches, P.M., Spagnuolo, E., Fernando, M., Pereda, P., Tarigo, A., and Verdier, V., (2010), Aneurysms of The Middle Cerebral Artery Proximal Segment (M1) • Anatomical and Therapeutic Considerations • Revision of A Series. Analysis of a series of the pre bifurcation segment aneurysms, *Asian Journal of Neurosurgery*, vol. 5, no. 2, pp. 57–63.

42. Kim, J.W., Kim, H., Kim, D.R., and Kang, H.I., (2014), The Sundt encircling clip as a vascular rescue: a case report and a review of repair methods for arterial tearing, *Journal of Korean Neurosurgical Society*, vol. 55, no. 6, pp. 353–356.

43. Tanriover, N., Kawashima, M., Rothon Jr., A., Ulm, A.J., and Mericle, R.A., (2003), Microsurgical anatomy of the early branches of the middle cerebral artery, *Journal of Neurosurgery*, vol. 98, pp. 1277–1290.

44. Ulm, A., Fautheree, G., Turniover, N., Russo, A., Albaneese, E., Rothon Jr., A., Mericle, and Lewis, S.B., (2008), Microsurgical and angiographic anatomy of the middle cerebral artery aneurysms: prevalence and significance of early branch aneurysm, *Neurosurgery*, vol. 62, pp. 344–353.

45. PhilippN – Modification of http://commons.wikimedia.org/wiki/Image:Endot racheal_tube_inserted.png, Public Domain, https://commons.wikimedia.org/ w/index.php?curid=4035901

46. Qudaisat, I.Y., (2008), Coanda effect as an explanation for unequal ventilation of the lungs in an intubated patient, *British Journal of Anaesthesia*, vol. 100, no. 6, pp. 859–860.

47. By Wapcaplet – Own work, CC BY-SA 3.0, https://commons.wikimedia.org/w/ index.php?curid=830253

48. Fred, L.A., Levi, D., Levine, R.A., Larson, M.G., Evans, J.C., Fuller, D.L., Lehman, B., and Benjamin, E.J., (1999), Prevalence and clinical outcome of mitral valve prolapse, *New England Journal of Medicine*, vol. 341, no. 1, pp. 1–7.

49. Delling, F.N., and Vasan, R.S., (2014), Epidemology and pathophysiology of mitral valve prolapse: new insights into disease progression, genetics, and molecular basis, *Circulation*, vol. 129, no. 21, pp. 2158–2170.

50. Singh, R.G., Cappucci, R., Kramer-Fox, R., Roaman, M.J.m Kligfield, P., Borer, J.S., Kchreiter, C., Isom, W., and Devereux, R., (2000), Severe mitral regurgita- tion due to mitral valve prolapse: risk factors for development, progression, and the need for mitral valve surgery, *American Journal of Cardiology*, vol. 85, no. 2, pp. 193–198.

51. Derover, C., Magne, J., Moonen, M., Le Goof, Dupont, L., Hulin, A., Radermecker, M., Colige, A., Cavalier, E., Kolh, P., Pierard, L., Lancellotti, P., Merville, M.P., and Fillet, M., (2015), Biomarker for primary mitral regurgitation, *Clinical Proteomics*, vol. 12, Article ID 25.

52. Sarano, M., Avierinos, J.F., Zeitoun, D., Detaint, D., Capps, M., Nkomo, V., Scott, C., Hartzell, M.S., Schaff, V., and Tajik, A.J., (2005), Quantitative determinants

for the outcome of asymptomatic mitral regurgitation, *The New England Journal of Medicine*, vol. 352, pp. 875–83.

53. Grigioni, F., Sarano, M., Zehr, K.J., Baile, K.R., and Tajik, A.J., (2001), Ischemic mitral regurgitation: long-term outcome and prognostic implications with quantitative Doppler assessment, *Circulation*, vol. 103, pp. 1759–1764.

54. Grigioni, F., Detaint, D., Avierinos, J.F., Scott, C., Tajik, J., and Sarano, M.E., (2005), Contribution of ischemic mitral regurgitation to congestive heart failure after myocardial infarction, *Journal of American College of Cardiology*, vol. 45, pp. 260–277.

55. Suri, R.M., Clavel, M.A., Schaff, H.V., Michelena, H.I., Huebner, M., Nishimura, R.A., and Sarano, M.E., (2016), Effect of recurrent mitral regurgitation following degenerative mitral valve repair: long-term analysis of competing outcomes, *Journal of American College of Cardiology*, vol. 67, pp. 488–498.

56. De Bonis, M., et al, (2016), Surgical and interventional management of mitral valve regurgitation: a position statement from the European Society of Cardiology Working Groups on Cardiovascular Surgery and Valvular Heart Disease, *European Heart Journal*, vol. 37, no. 2, pp. 133–139.

57. Srinivasacharya, D., and Srikanth, D., (2012), Steady streaming effect on the flow of a couple stress fluid through a constricted annulus, *Archives of Mechanics*, vol. 64, no. 2, pp. 137–152.

58. Hayat, T., Iqbal, M., Yasmin, H., and Alsaadi, F., (2014), Hall effects on peristaltic flow of couple stress fluid in an inclined asymmetric channel, *International Journal of Biomathematics*, vol. 7, no. 8, pp. 1–34.

59. Srivastava, N., (2014), Analysis of flow characteristics of the blood flowing through an inclined tapered porous artery with mild stenosis under the influence of an inclined magnetic field, *Journal of Biophysics, Journal of Biophysics*, vol. 2014, Article ID 797142.

60. Lancellotti, P., et al, (2010), European Association of Echocardiography recommendations for the assessment of valvular regurgitation. Part 2: mitral and tricuspid regurgitation (native valve disease), *European Journal of Echocardiography*, vol. 11, pp. 307–322.

61. Stokes, V.K., (1966), Couple stresses in fluids, *Physics of Fluids*, vol. 9, pp. 1709–1715.

62. Jaw, R.L., (1997), Effects of couple stresses on the lubrication of finite journal bearings, *Wear*, vol. 206, pp. 171–178.

63. Sobh, A.M., (2008), Interaction of couple stresses and slip flow on peristaltic transport in uniform and nonuniform channels, *Turkish Journal of Engineering and Environmental Sciences*, vol. 32, pp. 117–123.

64. Srinivasacharya, D., and Srikanth, D., (2008), Effect of couple stresses on the pulsatile flow through a constricted annulus, *C.R. Mecanique*, vol. 336, pp. 820–827.

65. Kolin, A., (1936), An electromagnetic flow meter: principle of the method and its application to blood flow measurements, *Experimental Biology and Medicine*, vol. 35, pp. 53–56.

66. Abd elmaboud, Y., Mekheimer, K.S., and Abdellateef, A.I., (2013), Thermal properties of couple-stress fluid flow in an asymmetric channel with peristalsis, *Journal of Heat Transfer*, vol. 135, pp. 1–8.

67. Rathod, V.P., and Tanveer, S., (2009), Pulsatile flow of couple stress fluid through a porous medium with periodic body acceleration and magnetic field, *Bulletin of the Malaysian Mathematical Sciences Society*, vol. 32, no. 2, pp. 245–259.

68. Sahu, M.K., Sharma, S.K., and Agrawal, A.K., (2010), Study of arterial blood flow in stenosed vessel using non-Newtonian couple stress fluid model, *International Journal of Dynamics of Fluids*, vol. 6, no. 2, pp. 248–257.

69. Korchevskii, E.M., and Marochnik, L.S., (1965), Magnetohydrodynamic version of movement of blood, Biophysics (Oxford), vol. 10, pp. 411–413.

70. Tzirtzilakis, E.E., (2005), A mathematical model for blood flow in magnetic field, *Physics of Fluids*, vol. 17, pp. 1–15.

71. Alimohamadi, H., Imani, M., and Forouzandeh, B., (2015), Computational analysis of transient non-Newtonian blood flow in magnetic targeting drug delivery in stenosed carotid bifurcation artery, *International Journal of Fluid Mechanics Research*, vol. 42, no. 2, pp. 149–169.

72. Taylor, C.A., Hughes, T.J.R., and Zarins, C.K., (1998), Finite element modeling of a blood flow in arteries, *Computational Methods in Applied Mechanics and Engineering*, vol. 158, pp. 155–196.

73. Formaggia, L., Lamponi, D., and Quarteroni, A., (2003), One dimensional models for blood flow in arteries, *Journal of Engineering Mathematics*, vol. 47, pp. 251–276.

74. Canic, S., Hartley, C.J., Rosenstrauch, D., Tambaca, J., Guidoboni, G., and Mikelic, A. (2006), Blood flow in compliant arteries: an effective viscoelastic reduced model numerics and experimental validation, *Annals of Biomedical Engineering*, vol. 34, pp. 575–592.

75. Jarez, S., and Uh, M. (2010), A flux-limiter method for modeling blood flow in the aorta artery, *Mathematics and Computer Modelling*, vol. 52, nos. 7–8, pp. 962–968.

76. Quarteroni, A., Tuveri, M., and Veneziani, A., (2000), Computational vascular fluid dynamics problems, models and methods, *Computing and Visualisation in Science*, vol. 2, pp. 163–197.

77. Rideout, V., and Dick, D., (1967), Difference-differential equations for fluid flow in distensible tubes, *IEEE Transactions of Biomedical Engineering*, BME, vol. 14, pp. 171–177.

78. Moffatt, H., (1964), Viscous and resistive eddies near a sharp corner, *Journal of Fluid Mechanics*, vol. 18, pp. 1–18.

79. Sobey, I., and Drazin, P., (1986), Bifurcations of two-dimensional channel flows, *Journal of Fluid Mechanics*, vol. 171, pp. 263–287.

80. Fearn, R., Mullin, T., and Cliffe, K., (1990), Nonlinear flow phenomena in a symmetric sudden expansion, *Journal of Fluid Mechanics*, vol. 211, pp. 595–608.

81. Hawa T., and Rusak, Z., (2001), The dynamics of a laminar flow in a symmetric channel with a sudden expansion, *Journal of Fluid Mechanics*, vol. 436, pp. 283–320.

82. Dijkstra, H., Wubs, F., Cliffe, A., Doedel, E., Dragomirescu, I., Eckhardt, B., Gelfgat, E., Hazel, A., Lucarini, V., Salinger, A., Phipps, E., Sanchez-Umbria, J., Schuttelaars, H., Tuckerman, L., and Thiele, U., (2014), Numerical bifurcation methods and their application to fluid dynamics: analysis beyond simulation, *Communications in Computational Physics*, vol. 15, no. 1, pp. 1–45.

83. Quaini, A., Glowinski, R., and Canic, S., (2016), Symmetry breaking and preliminary results about a Hopf bifurcation for incompressible viscous flow in an expansion channel, *International Journal of Computational Fluid Dynamics*, vol. 30, no. 1, pp. 7–19.

84. Chao, K., Moises, V., Shandas, R., Elkadi, T., Sahn, D., and Weintraub, R., (1992), Influence of the Coanda effect on color Doppler jet area and color encoding, *Circulation*, vol. 85, pp. 333–341.
85. Drikakis, D., (1997), Bifurcation phenomena in incompressible sudden expansion flows, *Physics of Fluids*, vol. 9, pp. 76–87.
86. Revuelta, A., (2005), On the two-dimensional flow in a sudden expansion with large expansion ratios, *Physics of Fluids*, vol. 17, no. 1, pp. 1–4.
87. Mishra, S., and Jayaraman, K., (2002), Asymmetric flows in planar symmetric channels with large expansion ratios, *International Journal of Numerical Methods in Fluids*, vol. 38, pp. 945–962.
88. Wille, R., and Fernholz, H., (1965), Report on the first European mechanics colloquium on Coanda effect, *Journal of Fluid Mechanics*, vol. 23, pp. 801–819.
89. Pitton, G., Quaini, A., and Rozza, G., (2017), Computational reduction strategies for the detection of steady bifurcations in incompressible fluid dynamics: applications to Coanda effect in cardiology, *Journal of Computational Physics*, vol. 344, pp. 534–557.
90. Oliveira, M., Rodd, L., McKinley, G., and Alves, M., (2008), Simulations of extensional flow in micro-rheometric devices, *Microfluidics and Nanofluid*, vol. 5, pp. 809–826.
91. Cherdron, W., Durst, F., and Whitelaw, J., (1978), Asymmetric flows and instabilities in symmetric ducts with sudden expansions, *Journal of Fluid Mechanics*, vol. 84, pp. 13–31.
92. Chiang, T., Sheu, T.W., and Wang, S., (2000), Side wall effects on the structure of laminar flow over a plane-symmetric sudden expansion, *Computational Fluids*, vol. 29, no. 5, pp. 467–492.
93. Tsai, C.H., Chen, H.T., Wang, Y.N., Lin, C.H., and Fu, L.M., (2007), Capabilities and limitations of 2-dimensional and 3-dimensional numerical methods in modeling the fluid flow in sudden expansion microchannels, *Microfluidics and Nanofluid*, vol. 3, no. 1, pp. 3–18.
94. Lauga, E., Stroock, A.D., and Stone, H.A., (2004), Three-dimensional flows in slowly varying planar geometries, *Physics of Fluids*, vol. 16, no. 8, pp. 3051–3062.
95. Wang, Y., Quaini, A., Canic, S., Vukicevic, M., and Little, S., (2016), 3D experimental and computational analysis of eccentric mitral regurgitant jets in a mock imaging heart chamber, NA & SC Preprint series no. 55, Department of Mathematics, University of Houston, http://www.uh.edu/nsm/math/research/NASC-preprint-series/NASC-preprint-series_pdfs/Preprint_No16-552.pdf
96. Van Houtte E, Van Lierde K, and D'Haeseleer E, et al, (2010), The prevalence of laryngeal pathology in a treatment-seeking population with dysphonia, *Laryngoscope*, vol. 120, pp. 306–312.
97. Barkmeier-Kraemer, J.M., and Patel R.R., (2016), The next 10 years in voice evaluation and treatment, *Seminars in Speech Language*, vol. 37, pp. 158–165.
98. Lowell, S.Y., Kelley, R.T., and Colton, R.H., et al, (2012), Position of the hyoid and larynx in people with muscle tension dysphonia, *Laryngoscope*, vol. 122, pp. 370–377.
99. By Anatomist90 – Own work, CC BY-SA 3.0, https://commons.wikimedia.org/w/index.php?curid=19678691
100. Titze, I.R., and Martin, D.W., (1999), Principles of voice production, *The Journal of the Acoustical Society of America*, vol. 104, issue 3, p. 1148.

101. Titze, I.R., (2006), *The Myoelastic Aerodynamic Theory of Phonation*, National Center for Voice and Speech, Iowa City.
102. Lucero, J.C., (1995), The minimum lung pressure to sustain vocal fold oscillation, *Journal of the Acoustic Society of America*, vol. 98, pp. 779–784.
103. McKinney, J., (1994), *The Diagnosis and Correction of Vocal Faults*, Genovex Music Group.
104. Allery, C., Guerin, S., Hamdouni, A., and Sakout, A., (2004), Experimental and numerical POD study of the Coanda effect used to reduce self-sustained tones, *Mechanics Research Communications*, vol. 31, pp. 105–120.
105. Guo, C., and Scherer, R.C. (1994), Finite element simulation of glottal flow and pressure, *Journal of the Acoustics Society of America*, vol. 94, no. 2, pp. 688–700.
106. Erath, B.D., and Plesniak, M.W., (2006), The occurrence of the Coanda effect in pulsatile flow through static models of the human vocal folds, *Journal of the Acoustics Society of America*, vol. 120, no. 2, pp. 1000–1011.
107. Alleborn, N., Nandakumar, K., Raszillier, H., and Durst, F., (1997), Further contributions on the two-dimensional flow in a sudden expansion, *Journal of Fluid Mechanics*, vol. 330, pp. 169–188.
108. Stern, V., and Hussain, F., (2003), Effect of deceleration on jet instability, *Journal of Fluid Mechanics*, vol. 480, pp. 283–309.
109. Ishizaka, K., and Flanagan, J.L., (1972), Synthesis of voiced sounds from a two-mass model of the vocal cords, *Bell System Technical Journal*, vol. 51, pp. 1233–1268.
110. Story, B.H., and Titze, I.R., (1995), Voice simulation with a bodycover model of the vocal folds, *Journal of the Acoustics Society of America*, vol. 97, pp. 1249–1260.
111. Adachi, S., and Yu, J., (2005), Two-dimensional model of vocal fold vibration for sound synthesis of voice and soprano singing, *Journal of the Acoustics Society of America*, vol. 117, pp. 3213–3224.
112. Nomura, H., and Funada, T., (2007), Effects of the false vocal folds on sound generation by an unsteady glottal jet through rigid wall model of the larynx, *Acoustics Science & Technology*, vol. 28, no. 6, pp. 403–412.
113. Zhang, C., Zhao, W., Frankel, S.H., and Mongeau, L., (2002), Computational aeroacoustics of phonation, Part II: effects of flow parameters and ventricular folds, *Journal of the Acoustics Society of America*, vol. 112, pp. 2147–2154.
114. Pelorson, X., Hirschberg, A., Wijnands, A.P.J., and Bailliet, H., (1995), Description of the flow through in-vivo models of the glottis during phonation, *Acta Acustica*, vol. 3, pp. 191–202.
115. Alipour F., and Scherer, R.C., (2002), Pressure and velocity profiles in a static mechanical hemilarynx model, *Journal of the Acoustics Society of America*, vol. 112, pp. 2996–3003.
116. Iijima, H., Miki, N., and Nagai, N., (1992), Glottal impedance based on a finite element analysis of two-dimensional unsteady viscous flow in a static glottis, *IEEE Trans. Signal Process*, vol. 40, pp. 2125–2135.
117. Liljencrants, J., (1989), Numerical simulations of the glottal flow, STL-QPSR, 1/1989, pp. 69–74.
118. Shinwari, D., Scherer, R.C., DeWitt, K.J., and Afjeh, A.F., (2003), Flow visualization and pressure distributions in a model of the glottis with a symmetric and oblique divergent angle of 10 degrees, *Journal of the Acoustics Society of America*, vol. 113, pp. 487–497.

119. Erath, B.D., and Plesniak, M.W., (2010), An investigation of asymmetric flow features in a scaled-up driven model of the human vocal folds, *Experiments in Fluids*, vol. 49, pp. 131–146.
120. Neubauer, J., Zhang, C., Miraghaie, R., and David, A.B., Coherent structure of the near field flow in a self-oscillating physical model of the vocal folds, *Journal of the Acoustics Society of America*, vol. 121, no. 2, pp. 1102–1118.
121. Dreschsel, J.S., and Thomson, S.L., (2008), Influence of supraglottal structures on the glottal jet exiting a two-layer synthetic, self-oscillating vocal fold model, *Journal of the Acoustics Society of America*, vol. 123, no. 6, pp. 4434–4445.
122. Triep, M., Brücker, C., and Schroder, W., (2005), High-speed PIV measurements of the flow downstream of a dynamic mechanical model of the human vocal folds, *Experiments in Fluids*, vol. 39, pp. 232–245.
123. Hofmans, G.C.J., Groot, G., Ranucci, M., Graziani, G., and Hirschberg, A., (2003), Unsteady flow through in-vitro models of the glottis, *Journal of the Acoustics Society of America*, vol. 113, no. 3, pp. 1658–1675.
124. Scherer, R.C., Shinwari, D., Dewitt, K.J., Zheng, C., Kucinschi, B.R., and Afjeh, A.A., (2001), Intraglottal pressure profiles for a symmetric and oblique glottis with a divergence angle of 10 degree, *Journal of the Acoustics Society of America*, vol. 109, no. 4, pp. 1615–1130.
125. Hirt, C., Amsden, A., and Cook, J., (1974), An arbitrary Lagrangian-Eulerian computing method for all flow speeds, *Journal of Computational Physics*, vol. 14, pp. 227–253.
126. Hughes, T.J.R., Liu, W.K., and Zimmermann, T., (1981), Lagrangian-Eulerian finite element formulation for compressible flows, *Computer Methods in Applied Mechanics and Engineering*, vol. 29, pp. 329–349.
127. Wall, W.A., Gerstenberger, A., Gammnitzer, P., Forster, C., and Ramm, E., (2006), Large deformation fluid structure interaction- advances in ALE methods and new fixed grid approaches, in H.J. Burngartz, and M. Schafer (eds), *Fluid Structure Interaction: Modeling, Simulation, Optimization*, LNSCE, Springer.
128. Lee, L., and LeVeque, R.J., (2003), An immersed interface method for incompressible Navier-Stokes equations, *SIAM Journal on Scientific Computing*, vol. 25, no. 3, pp. 832–856.
129. Mittal, R., and Iaccarino, L., (2005), An immersed boundary method, Annual Review of Fluid Mechanics, vol. 37, no. 1, pp. 239–261.
130. Peskin, C.S., (2002), An immersed boundary method, *Acta Numerica*, vol. 11, pp. 479–517.
131. Gerstenberger, A., and Wall, W.A., (2008), An extended finite element method/ Lagrange multiplier based approach for fluid-structure interaction, *Computer Methods in Applied Mechanics and Engineering*, vol. 197, pp. 1699–1714.
132. Alipur, F., Berry, D.A., and Ritze, I.R., (2000), A finite element model of vocal fold vibration, *Journal of the Acoustics Society of America*, vol. 108, pp. 3003–3012.
133. Thompson, S.L., Mongeau, L., and Frankel, S.H., (2003), Physical and numerical flow excited vocal fold models, in 3rd International Workshop, MAVEBA, Firenze University Press, pp. 147–150.
134. Tao, C., Jiang, J.J., and Zhang, Y., (2003), Simulation of vocal fold impact pressures with a self-oscillating finite element model, *Journal of the Acoustics Society of America*, vol. 119, no. 6, pp. 3987–3994.

135. Tao, C., Zhang, Y., Hottinger, D.C., and Jiang, J.J., (2007), An asymmetric airflow and vibration induced by the Coanda effect in a symmetric model of the vocal folds, *Journal of the Acoustics Society of America*, vol. 122, pp. 2270–2278.
136. Luo, H., Mittal, R., Zhang, Y., Bielamowize, S.A., Walsh, R.J., and Hahn, J. K., (2008), An immersed-boundary method for flow-structure interaction in biological systems with applications to phonation, *Journal of Computational Physics*, vol. 227, pp. 9303–9332.
137. Rosa, M.O., Pereira, J.C., Grellet, M.M., and Alwan, A., (2003), A contribution to simulating a three-dimensional larynx model using the finite element method, *Journal of the Acoustics Society of America*, vol. 114, no. 5, pp. 2893–2905.
138. Zhao, W., Franke, S.H., and Mongeau, L., (2001), Numerical simulation of sound from confined pulsating axisymmetric jets, *AIAA Journal*, vol. 39, no. 10, pp. 1868–1874.
139. Zhao, W., Zhang, C., Franke, S.H., and Mongeau, L., (2002), Computational aeroacoustics of phonation, Part I: computational methods of sound generation mechanisms, *Journal of the Acoustics Society of America*, vol. 112, pp. 2134–2146.
140. Zhang, Z., Mongeau, L., Franke, S.H., and Thompson, S.L., (2004), Sound generated by steady flow through glottis shaped orifices, *Journal of the Acoustics Society of America*, vol. 116, pp. 1720–1728.
141. Ffowcs-Williams, J.E., and Hawkins, D.L., (1969), Sound radiation by turbulence and surfaces in arbitrary motion, *Philosophical Transactions of the Royal Society London A*, vol. 264, pp. 321–342.
142. Crane, M.H., (2005), Aero-acoustic production of low frequency unvoiced speech sounds, *Journal of the Acoustics Society of America*, vol. 118, pp. 410–427.
143. Wagner, C., Huttl, T., and Sagut, P., (eds), (2007), *Large Eddy Simulation for Acoustics*, Cambridge University Press, New York.
144. Crighton, D.G., (1992), Computatinal aeroacoustics for low Mach number flows, in J.C. Hardin, and M.Y. Hussaini (eds), Computational Aeroacoustics, Springer-Verlag.
145. Lighthill, M.J., (1952), On sound generated aerodynamically. Part I: general theory, *Proceedings of the Royal Society London A*, vol. 21, pp. 564–587.
146. Lighthill, M.J., On sound generated aerodynamically. Part II: turbulence as a source of sound, *Proceedings of the Royal Society London A*, vol. 22, pp. 1–32.
147. Bogey, C., .Bailly, C., and Juve, D., (2002), Computation of flow noise using source terms in linearized Euler's equations, *AIAA Journal*, vol. 40, no. 2, pp. 1235–1243.
148. Oberai, A., Ronaldkin, F., and Hughes, T., (2002), Computation of trailing edge noise due to turbulent flow over an airfoil, *AIAA Aeroacoustic Journal*, vol. 40, pp. 2206–2216.
149. Ewert, R., and Schroder, W., (2003), Acoustic perturbation equation based on flow decomposition via source filtering, *Journal of Computational Physics*, vol. 188, no. 2, pp. 365–398.
150. Seo, J. and Moon, Y., (2006), Linearized perturbed compressible equations for low Mach number acoustics, *Journal of Computational Physics*, vol. 218, no. 2, pp. 702–719.
151. Kaltenbacher, M., Escobar, M., Becker, S., and Ali, I., (2008), Computational aeroacoustics based on Lighthill's analogy, in S. Marburg, and B. Nolte (eds), *Computational Acoustics of Noise Propagation in Fluids*, Springer.

152. Zorner, S., Kaltenbacher, M., and Dollinger, M., (2013), Investigation of pre-scribed movement in fluid–structure interaction simulation for the human phonation process, *Computers & Fluids*, vol. 86, pp. 133–140.
153. Link, G., Kaltenbacher, M., Breuer, M., and Dollinger, M., (2009), A 2D finite-element scheme for fluid-solid-acoustic interactions and its applications to human phonation, *Computational Methods in Applied Mechanical Engineering*, vol. 198, pp. 3321–3334.
154. Alipour, F., and Scherer, R.C., (2004), Flow separation in a computational oscil-lating vocal flow fold model, *Journal of Acoustics Society of America*, vol. 116, pp. 1710–1719.
155. Dollinger, M., and Berry, D.A., (2006), Visualization and quantification of the medial surface dynamics of an excised human vocal fold during phonation, *Journal of Voice*, vol. 20, no. 3, pp. 401–413.

Supplemental Reading List

Abbott, I.H., and von Doenhoff, A.E. (1956), *Theory of Wing Sections*, Dover Publications, New York.

Adrian, R.J., and Westerweel, J. (2011), *Particle Image Velocimetry*, Cambridge University Press.

Anderson, J.D. (2003), *Modern Compressible Flow: with Historical Perspective*, 3rd edition, McGraw-Hill Publishers, Boston.

Anderson, Jr., J.D. (2005), *Fundamentals of Aerodynamics*, 4th edition, McGraw-Hill Series in Aeronautical and Aerospace Engineering.

Applications of Circulation Control Technologies, Joslin, R.D., and Jones, G.S., (editors), Progress in Astronautics and Aeronautics, AIAA series

Batchelor, G.K. (2000), *An Introduction to Fluid Mechanics*, Cambridge University Press, 615.

Birkhoff, G., and Zarantonello, E.H. (1957), *Jets, Wakes, and Cavities*, Academic Press.

Bradshaw, P. (1971), *An Introduction to Turbulence and Its Measurement*, Pergamon Press, London.

Clancy, L.J. (1979), *Aerodynamics*, Pitman.

Davidson's Principles and Practice of Medicine (2014), 22nd Edition, edited by B.R. Walker, N.R Colledge, S.H. Ralston MD, pub: Churchill Livingstone/Elsevier.

Drain, L.E. (1980), *The Laser Doppler Technique*, John Wiley & Sons.

Durst, F., Melling, A., and Whitelaw, J.H. (1976), *Principles and Practice of Laser Doppler Anemometry*, Academic Press, London.

Gad-el. Hak, M. (2000), *Flow Control-Passive, Active, and Reactive Flow Management*, 1st edition, Cambridge University Press, UK.

Geankoplis, C.J. (2003), *Transport Processes and Separation Process Principles, (Includes Unit Operations)*, 4th Edition. Prentice Hall Professional Technical Reference - Technology and Engineering.

Hinze, J.O. (1975), *Turbulence*, 2nd Edition, McGraw-Hill, New York.

Houghton, E.L., and Boswell, R.P. (1969), *Further Aerodynamics for Engineering Students*, 1st edition, Edward Arnold (Publishers) Ltd, London.

Küchemann, D., (1978), *The Aerodynamic Design of Aircraft*, Pergamon Press, UK

Kuptsov, L.P. (2001), *Einstein Rule, in Hazewinkel, Michiel, Encyclopedia of Mathematics*, Kluwer Academic Publishers.

Lamb, H. (1945), *Hydrodynamics*, 6th edition, Dover Press, New York.

McCormick, B.W. (1997), *Aerodynamics, Aeronautics, and Flight Mechanics*, 2nd Edition, John Wiley and Sons.

Melling, A. (1997), Tracer particles and seeding for particle image velocimetry, *Measurement Science and Technology*, vol. 8, no. 12, pp 1406–1416.

Panton, R.L. (1996), *Incompressible Flow*, 2nd edition, Wiley and Sons.

Pope, A., and Goin, K.L. (1999), *High-Speed Wind Tunnel Testing*, 3rd edition, John Wiley & Sons, New York.

Rae, W.H. Jr. and Pope A. (1984), *Low-Speed Wind Tunnel Testing*, 2nd edition, John Wiley & Sons, New York.

Reynolds, A.J. (1974), *Turbulent flows in Engineering*, John Wiley & Sons.

Roger Temam, R. (1984), *Navier–Stokes Equations: Theory and Numerical Analysis*, ACM Chelsea Publishing.

Sapiro, A.H. (1953), *The Dynamics and Thermodynamics of Compressible Fluid Flow*, John Wiley & Sons.

Schlichting, H., Gersten, K., Krause, E., Oertel, H. Jr., and Mayes, C. (2004), *Boundary-Layer Theory*, 8th edition, Springer.

Smits, A.J. (2014), *A Physical Introduction to Fluid Mechanics*, John Wiley and Sons, Inc.

Tennekes, H., and Lumley, J.L. (1972), *A First Course in Turbulence*, The MIT Press, Cambridge, Massassachusetts.

Walz, A., and Joerg, O.H. (1970), *Boundary Layers of Flow and Temperature*, The MIT Press, Cambridge, Massassachusetts.

White, F.M. (2006), *Viscous Fluid Flow*, McGraw-Hill.

Woods, L.C. (1961), *The Theory of Subsonic Plane Flow*, Cambridge University Press.

Index

Milton Keynes UK
Ingram Content Group UK Ltd.
UKHW031143141024
449569UK00024B/1107